Five Easy Pieces on Water

Renzo Rosso

Five Easy Pieces on Water

Essentials of Water Science explained by an Engineering Scholar

Renzo Rosso
Politecnico di Milano
Milan, Italy

ISBN 978-3-031-69275-8 ISBN 978-3-031-69276-5 (eBook)
https://doi.org/10.1007/978-3-031-69276-5

Cover Illustration: Sergio Fedriani, GUITAR, watercolor, 1996.

This Springer imprint is published by the registered company Springer Nature Switzerland AG
The registered company address is: Gewerbestrasse 11, 6330 Cham, Switzerland

Fresh drinking water is an issue of primary importance, since it is indispensable for human life and for supporting terrestrial and aquatic ecosystems. Sources of freshwater are necessary for health care, agriculture, and industry. Water supplies used to be relatively constant, but now in many places demand exceeds the sustainable supply, with dramatic consequences in the short and long term.

Encyclical Letter Laudato si'
of the Holy Father Francis, 2015

To Donatella,
to mother Aria,
to the two Riccardo in my life,
and to our unforgettable Rufus.

Preface

Water is the most essential substance for life on Earth. Water is a common good, a shared resource, a treasure trove of humanity. From a knowledge standpoint, water is not exclusive to engineers, physicists, mathematicians, or architects. The scholars of business administration, political or social sciences cannot claim a primacy on water. Not even the philosophers, although they were the first to investigate its nature. Water is the most shared natural resource of all living organisms and represents both the material and cultural legacy of humanity.

Having devoted my academic career to the study of water, I hope to leave behind a legacy that is kind, pleasant, and approachable while maintaining the highest standards of scientific rigor. In recent years, my classes have included not just aspiring engineers, but also architects, urban planners, landscape designers, and occasionally, art historians. Most of them were curious and passionate students without the mathematical toolkit of a physicist. In introducing them to the wonders of science, the mysteries, and the beauty of water, I had to forego formal mathematics. I have found that qualitative discussions, especially when bolstered by historical context, can elucidate numerous properties and behaviors of water on our planet. This even encompasses several fundamental principles that support our understanding of water.

Professional divisions undoubtedly serve practical purposes, yet they are not inscribed on the tablets of any natural or divine decree. The pursuit of understanding of water knows no bounds, as Vitruvius recognized. He was the father of all 'water masons', as a distinguished professor at the University of Padua used to call the hydraulic engineering scholars. Over the centuries, specialization has expanded humanity's technological repertoire significantly.

However, if reductionism becomes the sole approach to knowledge, our ability to truly observe nature diminishes to a mere glimmer.

The handouts in this booklet record five core lectures on water as delivered to the students of the Landscape Hydrology Studio at the Master in Landscape Architecture and Land Landscape Heritage of the Politecnico di Milano. In the transcript, I also took advantage of dialogues in the context of educational laboratories and flipped classrooms. Since the primary emphasis of these lectures was on the topic of water in its simplest form, I was freed from any strict disciplinary framework. A comprehensive grasp of water's fundamentals serves as a valuable resource intended for dissemination across rigorous scientific fields, practical technology applications, as well as social and human sciences domains.

The title of the booklet draws on the legacy of one of the most fascinating scientific works of the last century. I am speaking of Richard Feynman's *Six Easy Pieces*, published posthumously in 1994. A title perhaps chosen in association with *Five Easy Pieces*, the collection of four-handed piano pieces by the Russian composer Igor Stravinsky, published during World War I. The work of Feynman, one of the greatest physics scholars of the twentieth century, deals with classical lectures that bring the reader with the lightness of a balloon toward understanding fundamental topics such as the atom, basic physics, energy, gravitation, and the relationship of physics with other sciences. Indeed, they are the papers of the lectures that Feynman taught Caltech students from 1961 to 1963. He had imparted these essential concepts in a very straightforward manner, using primarily qualitative explanations and little formal mathematics.

I should have said at the outset that we are now mainly discussing freshwater. I hope, for the time being at least, that the inexperienced but inquisitive reader would peruse this book's pages objectively and with forgiveness. And I expect the knowledgeable reader to be kind and understanding.

Eugenio Pugliese Caratelli deserves my gratitude for his comprehensive discussion on many aspects of this book. I also thank Giovanni Vannucchi for suggesting incorporating key drawings and other amenities. Solomon Vimal shared his storytelling about Robert Horton, Paolo Frisi, and other pillars of water knowledge with me. David Butler is gratefully acknowledged for his help in improving my poor writing style in a foreign language.

Milan, Italy Renzo Rosso

Acknowledgments

I am grateful to all my students, especially those who openly challenged me and those who quietly criticized my lectures. I have received far more from them than I could have ever imagined and much less than I could offer them, since even the harshest criticism aids us in thinking critically, understanding, and growing.

Introduction

I fulfilled my graduation in hydraulic engineering in two steps, a review essay and a research dissertation. First I presented a short review of *The limits to Growth,*[1] a book that greatly influenced my professional life. Then, I discussed my work on extreme value prediction in hydrology.

There was significant debate regarding the graduation panel's consideration of a novel subject in civil engineering, but it was settled peacefully. The core was the research dissertation on flood hydrology, which involved a significant amount of effort to advance probability theory by an infinitesimal amount. Its application to the statistical prediction of flood discharge in the Magra River, a minor creek in northwestern Italy, did not exactly excite the panel, which was between bored and skeptical.

At the time, hydrology was viewed as a sort of hydraulics for beginners, or even worse. A distinguished English scholar dubbed a hydrology textbook that was recently released by an Irish colleague as "Hydraulics for Dummies." Throughout the 1970s, the most popular saying among Italian and European academics was "hydrology is to hydraulics like astrology is to astronomy."

Everything has changed. After fifty years, Hydraulics is not nearly as popular, or practiced in colleges and research facilities worldwide as Hydrology. Hydrologists were among the first to take the issue of global warming and its climatic effects seriously because the water cycle is the transmission belt of weather and climate. Currently, hydrology serves as the professional domain of choice for those working in engineering, planning, management, and water

[1] Meadows, D. H., Meadows, D. L., Randers, J., & Beherens III, W. W. (1972). *The limits to growth.* Fall Church: Potomac Associates.

economics. To the point that some malicious hydrologists consider hydraulics to be one of the numerous subfields of hydrological sciences.

Is it worth distinguishing? Does the knowledge of water still make sense if it is reduced, fragmented, or contained? Do watertight compartments still make sense if we acknowledge that we are all viewing water in all of its facets through the diverse lenses of our unique, both different and complementary mindsets?

"No" is my answer.

I concur with Edgar Morin that we must rediscover the significance of the qualitative approach in addition to the quantitative one. To this effect, we must connect rather than separate, integrate rather than reduce.[2]

This book aims to simplify things, although unfortunately, they are often a little more complicated than they seem at first glance. I have spent a significant portion of my life complicating problems, sometimes simple ones, and sometimes only to satisfy the craving for the beautiful mathematical form of knowledge. Therefore, I follow here the traditional method that aligns with the legacy of Leonardo da Vinci, who first observed nature before theorizing.

This book does not ignore complexity. We should never forget that oversimplifications and false assertions must be out-lawed, as claimed by Archimedes of Syracuse, the man who gave the Western world its first knowledge of water and, unbeknownst to him, a hydraulic engineer.

> Those who claim to discover everything but produce no proofs of the same may be confuted as having actually pretended to discover the impossible.[3]

No discipline or scholar is a master of the science of water, despite the fact that many academics in today's highly fragmented university fight for supremacy. Even though water is arguably the most studied scientific topic during the past three millennia, everyone should help expand on the many aspects that are yet unknown. Professional issues make sense in many practical situations, but they should not mask the transdisciplinary nature of water knowledge. Water knowledge is a field that has no bounds in its quest for knowledge.

The path of those who study the water cycle and those who design water turbines was divided by disciplinary fragmentation beginning in the second half of the nineteenth century. The fates of scholars who grasp theories and sophisticated calculus have gradually diverged from those of engineers who build aqueducts and sewers. Let alone the forking paths taken by the experts in water governance and the water architects.

---

[2] Morin, E. (1986). *La méthode III. La connaissance de la connaissance*. Paris: Seuil (*The method III. Knowledge of knowledge*, in French).

[3] Archimedes, *On spirals*, around 225 BC (as translated by T.L. Heath, The Works of Archimedes, Cambridge: At the University Press, 1897).

Reductionism has had a significant impact on science, engineering, and architecture. Although it sparked a lot of advances in human knowledge, it is not the sole method for investigating nature or fostering innovation. A river network is greater than the sum of its creeks, as well as the forest is greater than the sum of its trees,[4] because each creek or tree contributes to river or forest health just as every member of society contributes to the well-being of their communities.

Theory and practice can go together. For instance, Luis van Wittel, born in Naples, is regarded as one of the finest architects of the eighteenth century. The world's most exquisite buildings and gardens, the Royal Palace of Caserta, is his creation. But we also owe him one of the most significant aqueducts of the contemporary era. The magnificent Carolino Aqueduct was opened in 1762 to supply a huge amount of water to the Palace and its surrounding gardens. It features bridges that cross mountain ridges, tracking the proper slope for water conveyance, as well as the first iron ducts made in Calabrian foundries. It is no accident that the elderly Vanvitelli was summoned to Genoa in 1771 to resolve a long-standing political and economic conflict around the 1000-year-old Genoa civic aqueduct.[5]

Why am I calling upon the past?

Models and data have exact bounds. A colleague who devoted his life to hydrological measurements once whispered that "no one believes in models except those who implemented them; and everyone believes in data except the one who measured them." Using only the power of poetry, no data nor computations, Dante expertly describes the crucial processes in the Arno River that produce flood hazard in Florence, as well as the city's vulnerability.[6]

The paradigm of knowledge has been almost entirely quantitative over the past 200 years. The numbers pretty much say it all. The remainder is anarchy. Although as is frequently the case, I concur with Edgar Morin again when he states that

> calculation (statistics, surveys, GDP growth rate) conquers everything. Quantity drives out quality. Humanism is in regression under the technical-economic pressure.[7]

[4] Wohlleben, P. (2016). *The hidden life of trees: What they feel, how they communicate.* Vancouver: Greystone Book.

[5] Podestà, F. (1879). (*L'acquedotto di Genova 1071–1879*, Genova: Tipografia del Regio Istituto Sordo-Muti (*The Genoa aqueduct from 1071 to 1879*, in Italian).

[6] Dante Alighieri, *Divina Commedia, Purgatorio,* Canto V, 116–123.

[7] Morin, E. (2014). *Einseigner à vivre. Manifeste pour changer l'éducation.* Paris: Actes Sud/Play Bac.

I have no doubts that humanity needs to reevaluate the worth of qualitative knowledge relative to quantitative knowledge.[8] In keeping with this, I arranged here the handouts of five basic lectures that I delivered to my students according to the approach by Guelfo Pulci Doria who pioneered the now rediscovered need of understanding science via history and integrating human science with STEM knowledge.[9]

It is true that there is a big gap between creating knowledge and getting it out to the public, yet closing this gap is essential to society's progress. In the seventeenth and eighteenth centuries, during the Enlightenment, the idea of public science started to take shape. Knowledge was no longer only the purview of elites or privileged institutions, but rather was valued as a public good. Prominent scholars including Francis Bacon, Galileo Galilei, and Isaac Newton promoted the sharing of scientific information with the general public. Because of their conviction that the growth of research should benefit all people, organizations like scientific societies and universities were founded with the goal of encouraging public participation with science.

This concept changed over time, inspiring the creation of procedures and policies that support inclusivity, openness, and transparency in scientific research and public outreach. But this also made the distance between knowledge and its application wider. Many academics these days do not give a damn when it comes to giving simple lectures, speaking to the public in person, or utilizing cryptic language, ignoring direct communication in science. The waning confidence and interest in science in the third millennium is not unrelated to this mindset.

Public and scientific confidence can be increased through effective science communication. Researchers can fight disinformation and improve the credibility of their work by openly disclosing their techniques and findings. These lectures aim to close the knowledge gap between public science and various fields and stakeholders using water. They also provide basic information to a wider audience of individuals who are interested in learning more about water and bridging the gap between theory and practice.

Water is an essential but multiform substance. It is complex in its apparent simplicity, an odd mix of order, disorder and organization. The majority of real-world applications appear simple and intuitive, yet as formal knowledge has advanced, many water processes can only be explained by rather hard, non-linear mathematical principles. Mankind has been aware of this for

---

[8] Rosso, R. (2019). *The decline and renaissance of universities: Moving from the big brother university to the slow university*. Heidelberg: Springer.

[9] Pulci Doria, G. (2005). *Il corso di idraulica a partire dallo sviluppo storico-sociale della disciplina*, Napoli: Cuen (*A treaty of hydraulics moving from the historical and social developments of the discipline*. in Italian).

2300 years, since Archimedes of Syracuse pioneered the mechanical approach along with introducing surprising inventions such as the hydraulic screw and the water clock. He always used mathematics to prove his discoveries. In the realm of water, things are not always easy. Speaking of water, nothing is ever too easy.

I chose the book's contents to provide a short but comprehensive survey of water knowledge. This includes encouraging human sciences exploration among STEM students and piquing the interest of students in social, economic, and human sciences toward mysterious STEM fields. For mankind, comprehending *the nature of water* has been a constant challenge. The first step to exploring the water world is to describe the behavior of *water at rest.* Even though people have observed running water and have been moving water for thousands of years, there are still a number of unsolved problems with *water in motion.* Investigating climate and ecosystems, as well as designing and carrying out civil engineering projects, requires a thorough understanding of *the water cycle.* In the twenty-first century, one of the main barriers to humanity's peaceful coexistence and cooperation will be *the governance of water.*

Contents

1

The Nature of Water

Water is H_2O, hydrogen two parts, oxygen one,
but there is also a third thing, that makes it water
and nobody knows what it is.
David Herbert Lawrence, The third thing, in: Painses, *1929.*

"Cooking acts like a capricious nymph—often leading to despair. Yet, it also brings joy, for when you master a dish or conquer a challenge, you revel in the taste of victory." Pellegrino Artusi introduces his culinary masterpiece with this blend of irony, presenting it as the most captivating recipe book ever written.[1] The same introduction could describe water: very useful and humble and precious and chaste, echoing St. Francis of Assisi's teachings. However, water also possesses a mischievous nature when one delves into its characteristics, because the only common substance that can exist in the Earth's surface at the comparatively narrow range of temperatures and pressures as a gas, liquid, or solid naturally is water. At times, like in the case of some wintertime geyser explosions in Island, all three stages can even exist at the same location and time.

[1] Artusi, P. (1891). *La scienza in cucina e l'arte di mangiar bene*. Florence: Salvatore Landi (*Science of cooking and the art of good eating*, in Italian).

R. Rosso, *Five Easy Pieces on Water*, https://doi.org/10.1007/978-3-031-69276-5_1

Ideas of Water

People have developed ideas and images of water because it has always been essential to human existence and social development. These range from the ancient Sumerian myths, in which Enki brings order and fecundity to the earth by pouring water into the beds of the Tigris and Euphrates, to modern Armageddon visions, in which melting glaciers turn into flowing water, submerging entire civilizations.[2]

Throughout history, humankind has viewed water as the cornerstone of the cosmos and life itself. Ancient philosophers regarded water as the foundational element of nature, the substance within which things exist and ultimately decay. This reverence for water's spiritual significance is echoed in both classical Western and Indian Vedic mythologies, as well as biblical narratives, which state, "he who is not born of water and the Spirit cannot enter the kingdom of God" (John 3:5). In the Far East, the perception of water was slightly different; it was associated with darkness and the Moon, contrasting with light and the Sun, representing a form of cosmological dualism.

For Thales of Miletus, water was the essence of matter, the primary principle (Greek: arché, ἀρχή) of the cosmos. He believed that nature defines the birth of all things. Drawing from his observations of sea fossils, Xenophanes theorized water as one of the fundamental cosmic elements, alongside the concept of wet and dry. Heraclitus, on the other hand, maintained that everything is in flux, positioning water as the source and conclusion of all. Beyond the realm of natural philosophy, the poet Pindar, who was more concerned with the tangible than the abstract, regarded water as the most valuable resource.

While Thales believed water to be the sole substance, Empedocles expanded the list of basic substances to include not just water, but also fire, earth, and air—the latter being considered the primal substance by Anaximenes. Empedocles' doctrine of the four elements—Fire, Air, Water, and Earth—subsequently integrated into Aristotelian philosophy, forming the foundation of Western thought for two thousand years.

Empedocles' four elements not only symbolized the primary substances but also corresponded to states of aggregation: solid (earth), liquid (water), gaseous (air), and etheric (fire), akin to aether. These elements could self-organize in countless proportions through a mechanism Empedocles proposed to

[2]Tvedt, T., & Ostigaard, T. (2009). A history of the ideas of water: Deconstructing nature and constructing society. In: *Ideas of water. From ancient societies to the modern world*, edited by T. Tvedt and Terje Ostigard, London: I.B. Tauris.

explain the formation of composite substances. However, not all combinations were feasible: water could mix with wine but not with oil. This distinction raised significant metaphysical discussions, highlighting water's dual nature as both the essence of purity and a fundamental element for life on Earth.

Leucippus and Democritus introduced the concept of atomic theory, positing that substances are aggregates—varying in stability—of indivisible, indissoluble, eternal, and unchangeable elements known as atoms. They envisioned everything as composed of these particles, which are both motionless and infinitely numerous in any given substance. This revolutionary idea had the potential to unlock vast realms of physical knowledge, yet it didn't gain much traction in the centuries that followed, despite Aristotle's commendation:

> Leucippus and Democritus systematically elucidated the nature of things, largely agreeing in their theories. They proposed a principle that aligns closely with the true nature of the universe.[3]

At that time, atomic theory struggled to account for many phenomena. For instance, if water consisted of atoms, natural philosophers were at a loss to explain why salt dissolved in water while sand did not, or what kept water's atoms together. They also grappled with understanding why atoms formed tangible bodies, solids, and liquids, instead of merely drifting freely through space.

Plato was among the pioneers to delve into the geometry of water, focusing on two basic shapes: two types of right-angled pseudo-triangles. Utilizing these shapes, he constructed four regular solids to represent Empedocles' four elements within regular solid frameworks. Earth's atoms were depicted as cubes; Fire's as tetrahedra; Air's as octahedra; and Water's as icosahedra. Plato's geometric interpretation of nature was widely esteemed, discussed, and refined across subsequent centuries (Fig. 1.1). This theory remained highly influential until the Renaissance, when Leonardo da Vinci, some eighteen centuries later, offered a radical critique in his *Manuscript F*: "I respond to Plato that geometric figures do not possess the essence of the universe's elements as he believed."

For Aristotle, water was a fundamental component of the sub-lunar world, shaping the Earth's surface. Above it lay the realm of air, and even higher, the domain of fire, which drove the dynamics of fluids. Aristotle's perspective was fundamentally metaphysical, with the path to understanding rooted in differentiating between "form" and "matter." For instance, in constructing a

---

[3] Aristotele, *De Generatione et corruptione*, A, 8, 324 b.

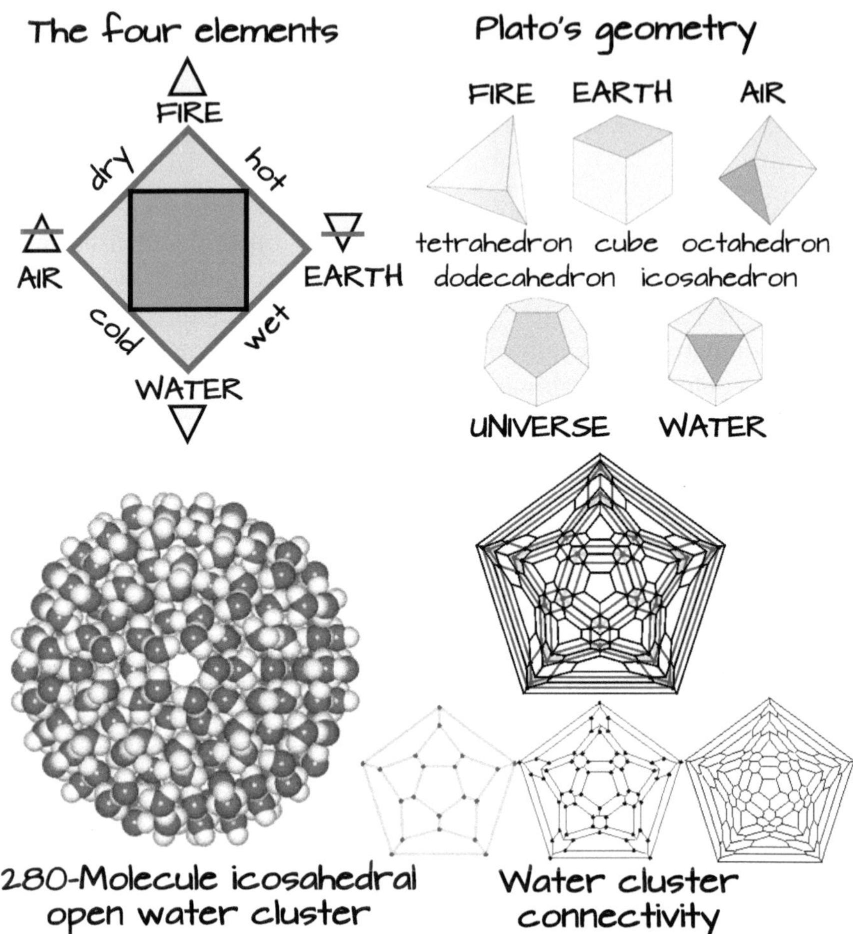

Fig. 1.1 Geometry of water

bronze sphere, the bronze serves as the matter, and its spherical shape represents the form. Similarly, in the calmness of a quiet sea, the water constitutes the matter, while its tranquility is the form.

Aristotle theorized that the four elements—Water, Air, Earth, and Fire—pair off to embody the four fundamental qualities: Hot, Cold, Wet, and Dry. As these elements warm up or cool down, they undergo transformations, swapping one or occasionally two of these qualities. For example, transitioning from cold to hot transforms water into air. These elemental changes, driven by the shifting seasons and thus governed by celestial movements, dictate the destiny of the Earth.

The enduring cultural influence of Aristotelian thought on water is evident in the themes explored in contemporary deep fiction, such as Hesse's *Siddhartha*, where the ferryman and the protagonist

> both listened to the water, which was no water to them, but the voice of life, the voice of what exists, of what is eternally taking shape.[4]

This long-lasting heritage has shaped the philosophical and scientific thinking of Western countries for over two millennia, significantly impacting theology as well.

Philosophy and poetry frequently intersect in discussions about water. In the renowned Latin educational poem, De Rerum Natura, water is a central theme, portrayed in all its complexity. Lucretius employs a variety of terms for water: aqua, liqueur, fluo, and humor. Each term is selected with care, aiming to more accurately describe the different aspects and characteristics of the water element that Lucretius seeks to highlight at various points.[5]

The teachings of Democritus and Epicurus are vividly reflected in the lexicon of flows. By employing *fluo*, along with its compounds and derivatives, Lucretius consistently underscores the concept of flux and flow. This notion of flow emerges as a defining characteristic across the physical, natural, and human realms, aptly illustrated through the analogy between the movement of atoms and the flow of rivers.

Beyond philosophical speculation and poetic endeavors, humanity quickly encountered numerous practical issues, much like those dear to Pindar. For centuries, sailing was the safest mode of travel and trade, making water in seas, lakes, and rivers a critical element to manage. Ensuring access to clean water for human settlements became essential, along with protection against storms and floods. Consequently, the quantity and quality of freshwater have emerged as key concerns in the stewardship of this precious resource.

Vitruvius, regarded as the father of architectural science, advised that anyone looking to found a new city should steer clear of marshy locations, opting instead for sites close to abundant, pure, and reliable water sources. By his time, Rome was already serviced by seven aqueducts. Later, Emperor Nero constructed the three Subiaco dams, serving dual purposes as sites of leisure and as reservoirs designed to control the inflow to the newly built Anio Novus aqueduct by Emperor Claudius. The largest of these dams was not only the

---

[4] Hesse, H. (1922). *Siddhartha: Eine Indische Dichtung*. Berlin: Fisher (*Siddhartha: An Indian Novel*, New York: New Directions, 1951).

[5] Luzzi, R. (2010). *Splendor aquae: the lexicon of water in De rerum natura*. PhD dissertation. Napoli: Università Federico II (in Italian).

tallest in the Roman Empire but also in the western world at that time. It held this record until 1305 AD, when it was likely destroyed by a severe flood.

Vitruvius, who lived at the turn of the Christian era, grudgingly accepted the Greek lesson that

> all bodies are made up of elements that the Greeks call the stichia, which are Fire, Water, Earth and Air.[6]

Being a practical man, he addressed the properties of water while keeping his treatise *De Architectura* firmly anchored in the achievement of the building process's three main objectives: strength, utility, and beauty. But he accomplished more than that. While defending the spring of Salmacis from untrue tales, he also assessed the potential of water as a driving force and took into account both the quantity and quality of springs.

> On this spring runs the false rumor, which ties the venereal disease to those who drink it. One will not apologize, however, for hearing how that false rumor is propagated. Not only can it not be, as they say, that water has made somebody effeminate and immodest, but it is rather a clear water with excellent taste.[7]

Subsequently, Pliny the Elder collected several opinions on the kind of water that goes best with meals in his writings. Prior to then, one of the most well-known physicians in history, Galen of Pergamon, had researched the characteristics of the water that improved hygiene. Under his sycamore tree, which is still standing today on the island of Kos, Hippocrates connected the "phlegmatic" temperament, one of the four human temperaments, to water. His water-based medicine states that a person dominated by water is generally quiet, thoughtful, sensitive, submissive, faithful, and honest; nevertheless, if the water is dirty, he becomes boring and uninterested in an unbalanced scenario.

The myth of the Flood reveals the dark side of the moon, despite the fact that water has traditionally been perceived as the essential component of life and is viewed primarily positively. The unfavorable perception of water was associated with natural calamities such as floods and storms at sea. Furthermore, it was once thought that the impurity of water represented the filth of the human spirit. For instance, bathing was seen by Early Christians as a

[6] Galiani, B. (1790). *L'architettura di Marco Vitruvio Pollione tradotta e commentata dal marchese Berardo Galiani*, Vol. I (Cap. IV), Naples: Fratelli Terres (*The architecture of Marco Vitruvio Pollione, translation and comments by Marquis Berardo Galiani*, in Italian).

[7] Galiani. (1790). *ibidem*, B.II, C.VIII.

temptation rather than a fault as they considered temperance to be a cardinal virtue.

> Who knows what impure thoughts might arise in a tub of warm water? With this danger in mind, St. Benedict declared, "To those who are well, and especially to the young, bathing shall seldom be permitted." St. Agnes took the injunction to heart and died without ever bathing.[8]

It wasn't until the late Middle Ages that water was thought to have inherent negative qualities. For instance, water was considered to be a source of infection during the fourteenth-century epidemic known as the plague or the black death. It was believed to open skin pores and let the purported germs, known as seminaria, to enter the body. As a result, people were advised against washing because water was thought to weaken the body.

Nonetheless, until the latter decades of the eighteenth century, people held the view that water was a primitive and inseparable element. The nature of water was still largely unknown, despite the discoveries made by Galileo Galilei and Leonardo da Vinci, who both contributed to the scientific community and solved numerous puzzles. The jump occurred when scientists from France and England, including Cavendish, Priestley, and Lavoisier, revealed to humanity that this material is actually composed of two basic elements: hydrogen and oxygen. Democritus's theory demonstrated its prophetic usefulness.

The Final Cut on Water Composition

Specifically, what about the water? The solution is quite straightforward: a molecule made up of two hydrogen atoms and one oxygen atom bonded together. Antoine-Laurent de Lavoisier's laboratory experiments at the end of the eighteenth century demonstrate this. He showed that water is not a single element but rather is subject to decomposition and re-composition as a result of two gas components combining at room temperature and pressure. He conceived a clever method to divide water into these two components in order to demonstrate his conjecture. He blew red-hot water vapor over the iron of a cannon barrel. In this manner, iron and oxygen were mixed, allowing Lavoisier to collect the hydrogen in a glass bell. The discovery of water's breakdown

---

[8] Kelly, J. (2006). *The great mortality: An intimate history of the black death, the most devastating plague of all time*. Glasgow: Harper Perennial.

disproves Aristotle's theory of Four Elements once more and this discovery establishes that water is composed of the components oxygen and hydrogen.

Indeed, Lavoisier and his associates Bucquet and Laplace conducted more complex experiments. Lavoisier's apparatus for studying the reaction between water and iron at high temperature is described in his 1789 *Traité Élémentaire de Chimie*. Several members of the Royal Academy of Sciences attended the final experiment because the verification of this fact was very important to chemical and physical theories (Fig. 1.2).

Water steam was transferred to a high-temperature glass tube and reacted with a spiral-shaped iron sheet within by heating the distilled water inside the appropriate retort. Hydrogen and black iron oxide were produced when iron and water reacted. After traveling through a lengthy tube, the unreacted water steam condensed in the condenser and was gathered in a bottle. With the bell

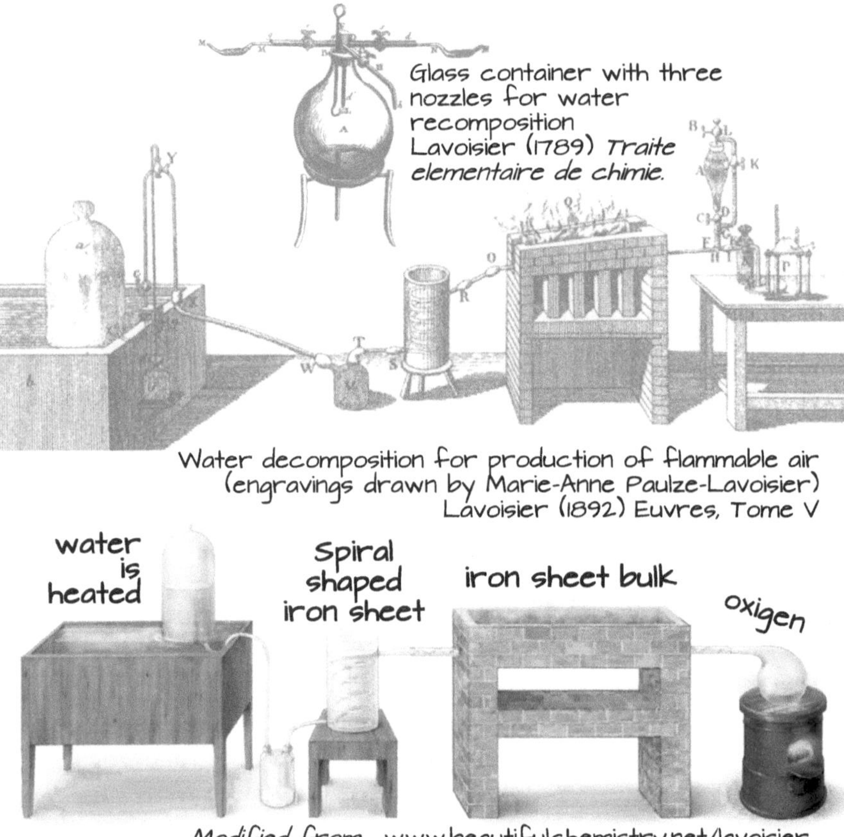

Fig. 1.2 Lavoisier's experiment

jar resting on a water trough's platform, hydrogen was collected, and its mass was evaluated. The oxygen in the water caused the iron sheet's bulk to grow. Lavoisier meticulously verified that the entire mass of hydrogen gathered in the bell jar and oxygen determined from the increase in iron weight matched the mass of reacted water (the mass loss of the retort subtracted from the mass increase of the collecting bottle). Using this cumbersome arrangement, Lavoisier discovered that the weight percentages of hydrogen and oxygen in water are 15% and 85%, respectively. This public experiment—performed by Lavoisier on 24th of June 1783 to demonstrate water decomposition for production of "flammable air"—pioneered public science.[9]

Lavoisier's discovery had been pioneered in a 1766 paper titled *On Factitious Airs* by British natural philosopher, theoretical chemist, and physicist Henry Cavendish. Like Newton, Boyle, and Lavoisier, he was a "scientific gentleman." Each of these scientists came from an aristocratic family and had enough money to construct their own laboratory at home.

Cavendish described the density of inflammable air and observed that as it burned, it formed water. This is regarded as the discovery of hydrogen because it establishes water as a compound rather than an element by identifying the gas as a distinct entity that, when interacted with oxygen, created water, referred to as "dephlogisticated air." Cavendish spoke in terms of the antiquated chemistry "phlogiston" idea. After a metal–acid interaction, he named the gas "flammable air." After repeating Cavendish's experiment, Antoine Lavoisier dubbed the element "hydrogen" after the Greek terms "hydro" and "genes," which combined mean "water forming."

Even though Lavoisier's discovery was ultimate, several members of the scientific community originally opposed and mistrusted his work. While some scientists doubted the veracity of Lavoisier's experiments and Cavendish's theories, others were reluctant to give up on the idea that water is an elemental substance. Amadeo Avogadro Count of Quaregna and Cerreto, an Italian gentleman scientist, did not discover the H_2O formula for water until 1811[10]:

$$2H_2 + O_2 \rightarrow 2H_2O$$

[9] Lehman, C., & Bensaude-Vincent, B. (2007). Public demonstrations of chemistry in eighteenth century France. *Science & Education, 16*, 573–583.

[10] Avogadro, A. (1811). Essai d'une manière de déterminer les masses relatives des molécules élémentaires des corps. *Journal de Physique, 73*, 58–76 (An attempt to determine the relative masses of the elementary molecules of bodies and the proportions by which they enter into these combinations, in French).

Avogadro's theory was not widely accepted until 1860, and chemists of the day occasionally provided distinct formulas to distinct substances. Three theories were most popular for water: Dalton's HO, Avogadro's H_2O, and other French chemists' H_4O_2. Water was ultimately identified as the component of the H_2O formula only after more methods and discoveries were made.

The discovery of water ionization—the process by which water molecules split into ions in an aqueous solution—was another important milestone. According to the equilibrium equation, a tiny percentage of water molecules in pure water spontaneously react to generate hydronium and hydroxide ions. The reaction

$$H_2O \rightarrow H^+ + OH^-$$

was introduced by 1903 Nobel prize winner Svante Arrhenius.[11] In developing a theory to explain the ice ages, he was the first to use in 1896 physical chemistry to estimate how much an increase in atmospheric carbon dioxide (CO_2) would raise Earth's surface temperature through the greenhouse effect.

In practice, we only discovered that water is a fairly simple substance in modern times. It is made up of two hydrogen atoms, which are the second most common element in the universe after helium, and one oxygen atom, which is the most plentiful element in nature.[12] But water's physical and chemical behavior has several characteristics that are so unique that they have sparked wild speculation among experts from antiquity as well as more contemporary times. In the latter part of the 1900s, for instance, there was a spirited discussion about the potential for water to solidify into a polymer known as polywater. This was a scenario that could have fatal consequences for all living things.

Of all the known fluids, Felix Franks claims that water has likely been studied the most and understood the least. Biophysicist Franks has developed novel techniques for delivering insulin and freezing vaccines. His work has changed scientific perceptions of water. Additionally, he placed the final touches on the hoax polywater discovery, which had a profound effect on world politics. This is the enigmatic Soviet finding that poisoned the Cold War and provided material for several spy and satirical novels.[13]

---

[11] Arrhenius, S. (1887). Über die Dissociation der in Wasser gelösten Stoffe. *Zeitschrift für Physikalische Chemie*, vol. 1U(1), 631–648 (On the dissociation of substances dissolved in water, in German).

[12] In units of mass.

[13] See. e.g., Vonnegut, K. (1964). *Cat's cradle*. New York: Holt, Rinehart and Winston; Myers, H. L. (1971). Polywater Doodle. *Analog Science Fiction and Fact,* February; O'Brian, R. C. (1972). *A report from group 17*. New York: McMillan.

The anxiety of deepening the knowledge of a liquid simple only in appearance has produced other fakes. For example, in the late 1980s of the twentieth century, the startling discovery of water memory gained immense popularity.[14] The scientific community was shocked by a French study team led by immunologist Jacques Benveniste when they revealed that liquid water can maintain a "footprint" of the materials it has come into contact with. Their research demonstrated that water possessed the ability to retain knowledge about things that had been diluted or dissolved, acting as a sort of memory. This might offer a strong defense of the concepts underpinning homeopathic medicine.

This theory caused a stir and led to heated discussions and violent arguments. The esteemed publication *Nature*, which had first reported the discovery, decided to resolve the debate by appointing a scientific committee to assess the discovery's validity and extent, above all. The dispute was resolved definitively, shutting down the matter once and for all. It was thought that the French researcher and his secretary had purposefully falsified the data by heavily manipulating the results, not to mention not being repeatable.[15] And the recollection of the memory of the water was quickly disappearing.

The Chemical Nature of Water

Why is the concept of water so poorly understood, and even misinterpreted at times?

Water exhibits distinct chemical behavior of its own. The temperatures at which it fuses and boils are anomalous. Since most life-related events take place at temperatures between 20 and 30 °C, it is these anomalies that guarantee life on Earth.

First, compared to its solid counterpart, ice, liquid water has a higher density. At 4 °C, water gets its highest density of one gram per cubic centimeter. Water solidifies below zero, reducing its density by roughly ten percent. As a result, ice solidifies and takes up more volume than the same quantity of liquid water. This allows ice to float on the water, much like icebergs, with just 10% of their bulk rising above the ocean's surface. There are additional repercussions. For instance, because of the way lakes freeze, life may exist there

[14] Davenas, E., Beauvais, F., Amara, J. et al. (1988). Human basophil degranulation triggered by very dilute antiserum against IgE. *Nature*, *333*, 816–818.

[15] Maddox, J., Randi, J., & Stewart, W. W. (1988). 'High-dilution' experiments a delusion. *Nature*, *334*(6180), 287–290.

even in extremely cold temperatures. Instead of starting at the bottom and working its way down, the freezing process begins at the surface.

Significant energy storage capability is implied by the high value of water thermal capacity. Because of this characteristic, aquatic bodies may store a lot of thermal energy. This characteristic allows the warm ocean currents—the Curoscivo Current in the Pacific, the Agulhas Current in the Indian Ocean, and the Gulf Stream in the Atlantic Ocean—to transfer massive amounts of heat from the ocean into the atmosphere. The reason that nearly the whole globe has a temperate climate, otherwise rendering it uninhabitable, is due to these massive currents.

Water is a cunning nymph because, even in stillness, it tantalizes the observer by revealing itself to be radically different depending on visual capabilities of the observer. As Richard Feynman used to tell his students,

> if you magnify a drop of water with an advanced microscope, it looks like a bunch of paramecia.[16]

Beautiful unicellular ciliated protozoa called paramecia are found in a variety of freshwater, brackish, and marine habitats. Because its cilia, which are organized in closely spaced rows around the outside of the body, force the paramecium forward, it is able to move and rotate.

Let us take a half-centimeter-diameter drop of water and examine it closely. You can get a 200× smartphone microscope lens for only 20 euros! Despite our best efforts, all we are able to see is water—smooth, unbroken water. Even at two thousand magnification, which is the highest power available for optical microscopes, we can only ever see reasonably smooth water.[17]

If one looks more closely, the microscope displays small, weird bodies that like soccer balls that are swimming here and there, from front to back. These bodies look like paramecia. Here, you can pause and admire the paramecia, their squiggling cilia, and their coiled bodies. You could even enlarge these amusing creatures a little bit in order to examine them more closely, much like biologists do. However, everything changes if we investigate that watery stuff in greater detail and magnify it by an additional two thousand times.[18] The object we are looking at becomes less smooth and appears as a swarming mob; it resembles the crowd seen from a very far distance away from the Woodstock Rock Festival venue. Drawn in by this formless throng, one magnifies the

---

[16] Feynman, R. P., Leighton, P., & Sands, M. (2005). *The Feynman lectures on physics: The definitive and extended edition, Chap. 1*, 2ª ed., Reading: Addison Wesley.

[17] The drop will be as large as a classroom, with a diameter of about 16 m.

[18] The drop will be about 20 km in diameter.

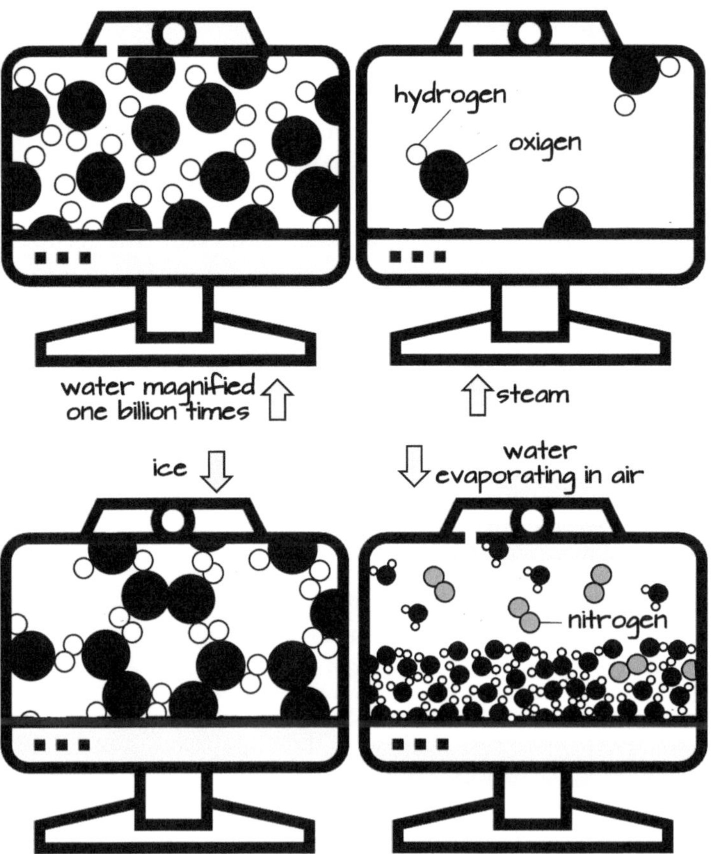

Fig. 1.3 Atomic geometry

picture an additional 225 times, seeing something odd that resembles a polka-dot silk tie from the distinguished Marinella of Naples.

Let us try harder, zooming in a billion times more (Fig. 1.3). We observe that the atoms of hydrogen and oxygen are quite different in size, forming two different types of blobs. A water molecule is formed when two hydrogen atoms, the twin little blobs, are connected by each giant oxygen blob. But take caution—the picture is out of focus, and the particles are always in motion—they bounce, flip, and stack one on top of the other. And we understand that we need to record a video clip because just one shot will not do. The water at rest never stays still on very small scales. If Leucippus and Democritus had realized this, they would have simply disregarded every argument made against them by their peers. Additionally, their view would have been more successful in protecting the world from Aristotle's enduring primacy in posterity.

Up until now, we have used the downscaling method demonstrated in the excellent short documentary *Powers of Ten*, which was created and filmed by a married couple of industrial designers, Charles and Ray Eames. This method shows how the universe is scaled relative to an order of magnitude or logarithmic scale.[19] It is not enough. The most important fact that cannot be revealed by a picture or a video clip is that the particles are attracted to one another and are stuck together; one is dragged by another, and so on. The group as a whole appears well glue together. The particles do not compress into one another at the same moment. They reject each other if someone attempts to squash two of them too closely together.

What is the behavior of this big drop of water with all these particles that are stuck together and oscillate?

First, because molecules are attracted to one another, water does not crumble or break into fragments; instead, it maintains its volume. A drop of water will flow, but it won't vanish, if it is placed on a slope where gravity might cause it to move from place to place. Furthermore, due to molecular attraction, it will remain unchanged a little bit further downstream, offering the last disproof of the Aristotelian dualism between matter and form. When Leonardo da Vinci pointed out that water can only go downward, his lesson was evident.

More precisely, what function does these particles' continuous oscillation serve?

The movement is what we interpret as heat; as the temperature rises, the oscillation increases. The volume between the atoms grows and oscillation becomes more intense when the water is heated. If the heating continues, the molecules will eventually drift apart because their attraction to one another is no longer strong enough to hold them together. As a result, we have reached the point at which liquid water turns into vapor, with the particles taking off as the temperature rises. When you cook "Spaghetti Alfredo" for dinner, you can observe that. Pellegrino Artusi was quite clear in his cooking handbook that as soon as the water starts to boil, the entire package of spaghetti needs to be placed into the pan.

It is easier to comprehend the appearance and nature of water when steam is observed rather than liquid water. Because the molecules are evenly spaced from one another, the molecule's structure seems distinct and unambiguous. An oxygen atom coupled to two hydrogen atoms forms a single molecule, with the two atoms forming an angle of roughly 105°. Atoms are 1 or 2 nm

---

[19] Charles and Ray Eames. (1977). *Powers of Ten*. Pyramid Films, based on Boeke, K. (1957). *Cosmic view: the Universe in 40 Jumps*. New York: John Day Company.

in size, or 1 to 2 billionths of a meter. Additionally, there is a tiny gap of less than 1 nm between the centers of the hydrogen and oxygen atoms.

The oxygen atoms of two water molecules interact with the hydrogen atoms of the other molecule as they get close to one another. The hydrogen bond, which is mostly electrical, is formed as a result. Its distinguishing feature is the existence of two single pairs of electrons positioned at a tetrahedron's two vertices, while hydrogen atoms occupy the other two vertices. The oxygen atom has a high electron charge thanks to these electron pairs, which allows oxygen to interact with the hydrogen atoms of two additional water molecules. Since the individual molecules of ice and liquid water are connected together, hydrogen bonds govern their structure. The water still has its own patterns even though these are far from the Plato icosahedra.

A lattice of interconnected molecules, dominated by tetrahedrons holding four water molecules, is formed with the energy of hydrogen bonds.[20] Phase transition temperatures are raised by the existence of hydrogen bonds because they contribute to an increase in the heat of evaporation and fusion.

Melting is the term used to describe the process of changing from a solid to a liquid. Rather, sublimation refers to the change from the solid to the gas phase. The change from a liquid to a gas is called evaporation. Condensation refers to the opposite occurrence, or the transition between the gaseous and liquid phases.

We are all familiar with melting. Every spring, as the snow melts in the sun and feeds the creeks through a thousand little streamlines, we are accustomed to watching this flow in the mountains. As I told above, solid water requires more room in the freezer than liquid water, as we learned when a glass bottle full of water had eventually broke in our refrigerator.

Perhaps best seen in the Great Lakes of North America, mist hanging over a lake is a breathtaking sight, although mist can be seen over any lake, especially in the fall. Because the air cools more quickly toward the end of summer than the water does, evaporation is the process that produces it. Warmer lakes cause a flow of warm, moist air to climb toward the colder, overlying mass of air when a mass of cold, dry air passes over them. According to the second law of thermodynamics, which states that when two bodies come into contact, the system tends to establish a steady state, the flow from the lake's surface to the air above quickly approaches a state of equilibrium. This process occurs just how Hippocrates would have liked—phlegmatically. Even though the vapor mist is only a thin layer, it is sufficient to block out sunlight, giving the surrounding area a surreal appearance, an eerie but magical look.

[20] About 5 kilocalorie per mole.

Because sublimation usually happens in certain climatic circumstances, including low relative humidity and dry winds, we are less familiar with it. At higher elevations, when atmospheric pressure is lower than in lowlands, snow and ice melt and become vapor. A large quantity of energy is needed for this process, and the sole source of that energy is bright sunlight. If I were to suggest a location on Earth to witness sublimation, I would choose the Everest's southern slope. Sublimation is best achieved in conditions of low air pressure, strong winds, bright sunshine, and low temperatures.

Conversely, de-sublimation—also referred to as icing or frosting—is the process by which a gas changes from a gaseous state to a solid one. This is the result of the vapor turning directly into ice without first entering the liquid phase. The most well-known instance of de-sublimation is likely the wintertime frost that forms on windows. In frigid air, water vapor never turns into liquid water—rather, it freezes into ice. This is also the process by which hoar frost formed, which explains why some frost occurs in home freezers. Weather conditions that have the potential to cause water ice to form on an aircraft's wings' surface are referred to as icing conditions in aviation.

A Mechanical Model of Water

Water is actually a cunning rascal because it changes drastically depending on the lens from which we view it. The first step in studying water, whether it is still or moving, is determining the appropriate spatial and temporal scales. Depending on whether a telescopic or microscopic magnifier is used to explore the features of the water, its nature appears to change. Additionally, this also depends on the time clock in use.

Quantum dynamics is essential at subatomic scales, on the order of billionths of a millimeter and millionths of a millisecond. Molecular dynamics is adequate at the atomic scale, which is ten-millionths of a millimeter and one millionth of a microsecond. We can examine water at rest or in motion from a tenth of a millimeter to a hundred milliseconds according to the classic physics of the continuum. On this scale, we study fluid mechanics. It is also the lowest hydraulic and hydrological reference scale available (Fig. 1.4).

When studying the behavior of liquid water, we do not analyze the behavior of the different molecules; but we replace the molecular structure of the substance with a hypothetical continuous medium, which at some point in space has the mean properties of the molecules surrounding that point. In practice, we are looking at a much wider radius sphere than molecular spacing. Thus, we call a water particle a small fluid element containing many

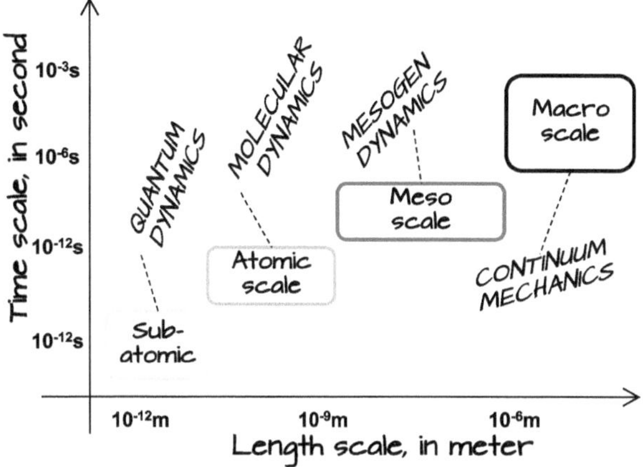

Fig. 1.4 Time and space scales

molecules and having the mean properties of the fluid at this particular position in space.

Liquid water has the ability to flow, or the ability to take on the shape of a vessel that it is contained in, much like all other fluids. For once, we are unable to refute Aristotle's distinction between form and matter. Furthermore, a fluid is a material that is unable to tolerate ongoing deformation. This indicates that, at the molecular level, two previously nearby liquid particles can be pushed apart by even a tiny force. Unlike solods, which present an elastic return, liquid particles do not have any tendency to approach one another once the cause of the deformation has gone. The various entities of the intermediate forces operating in the medium are linked to the differences in behavior between the liquid and the solid.

The "pesto" is a green sauce made with olive oil, crushed garlic, coarse salt, European pine nuts, basil leaves, and hard cheeses like Parmigiano-Reggiano and Sardinian Pecorino. The Italian region of Liguria's capital, Genoa, is where it all began. It is traditionally pounded in a marble mortar with a wooden pestle. An old Genoese proverb asks: what happens to those who pound water in the mortar? A totally pointless effort. Water can, in fact, resist severe compression forces. However, due to its extremely low resistance to traction forces, it may be pushed effortlessly. An analogy of smooth spheres can be used to visualize the movement of molecules in liquid water. We use this physical pattern for billiard balls, glass bead games, ball bearings, and plastic balls from children's playgrounds. These models explain the soft and hard nature of water as described by Lao-Tzu:

Nothing in the world
is as soft and yielding as water.
Yet for dissolving the hard and inflexible,
nothing can surpass it.[21]

Can this analogic model explain the nature of water better than these verses of a poem written twenty-third centuries ago?

An Insight of Solid Water

We are misguided if we believe we understand everything there is to know about water. So we return to the lab and examine water once more via the lens of an ever-stronger microscope. The mobility of molecules and atoms gradually slows down as the temperature drops when we cool our water drop. The swinging motion will eventually diminish and stop because atoms are attracted to one another and can no longer swing. The molecules condense into ice, a new solid structure, at very low temperatures. Every atom in this material has a precise location and forms a hard lattice that is easy to observe with the unaided eye. An ice needle remains securely attached to the rest of the needle if we raise it from one end while holding it in our hands. Compared to liquid water, which is fundamentally unstable due to atoms moving constantly, ice is substantially different.

Why does liquid water differ so much from solid water?

Atoms are not distributed randomly in a solid; instead, they are ordered in a crystalline matrix, which is a type of static lattice. Even across extremely long distances, the atoms maintain their cohesiveness and organization, forming a configuration akin to a Roman phalanx arrayed in a tortoise-like pattern. The locations of distant atoms on one face of the crystal, millions of atoms away, dictate the positions of the atoms on the opposite face of the crystal (Fig. 1.3 *bottom left*).

Every molecule in liquid water is far more free. It acts as if it has two arms and two focal points. However, it appears to have a will of its own because it settles in to promote tetrahedral coordination amongst the molecules. Thus, liquid water is composed of a dynamic lattice that is prone to deformation and exhibits numerous faults. As mentioned above, molecules occupy the interstitial spaces in the lattice, helping to enhance the density of liquid water in comparison to ice.

---

[21] Lao-Tzu (fourth–third century BC) *Tao Te Ching*, Room 78, translation by S. Mitchell, 1995.

Every oxygen atom in ice is joined to four other molecules by hydrogen bonds. By rotating the image on a 60-degree axis, you can observe the ice's structure: the image returns to its original state. This symmetry explains the snowflakes' six-sided shape.

Why does the ice shrink when it melts?

This specific crystalline model of ice includes several holes. Molecules easily fill these voids when the structure breaks. Most simple substances, with the exception of water and metals, expand when they melt, because the atoms in most simple substances are densely packed in their solid crystals. They require additional space to oscillate after melting, so they expand when melting, with the exception of metals and water. Instead, like an actually rascal nymph, an open structure like water collapses.

Ice is a hard crystal, nonetheless it has a temperature range. You can adjust the amount of heat in the ice if you'd like. Atoms swing and vibrate; they are not still. As a result, despite the crystal's precise geometric structure and well-defined order, every atom vibrates—atoms "move standing still," an oxymoron of nature. They vibrate with increasing amplitude as the temperature rises, until they start to shake and move. At such point, the ice melts. On the other hand, vibration decreases with decreasing temperature and keeps going down until atoms vibrate almost imperceptibly at absolute zero. In a nutshell, atoms never truly rest.

Because of the awkward nature of water, our beliefs start to falter when the temperature drops. At standard air pressure, the temperature drops to numerous condensed phases, including liquid and solid. Ice becomes vitreous or amorphous below −140 °C, and when heated, this phase transforms into an extremely viscous liquid. Ice crystallizes into its usual structure at temperatures higher than −120 °C. It is challenging to conduct experiments in the no man's land that is the area between −120 and −38 °C. And as we go closer to the border, we see ice crystals that change shape under increased pressure.

Knowledge about very cold water is still poor. Numerous aspects still need to be explored. Increasing our understanding of this field would lead to new discoveries in a variety of fields, including space exploration. The secrets of life and the universe are still largely shaped by the mysteries of water.

Water Sensing

If we assume that the art of cooking and the study of water are closely related fields, knowledge of water's taste, fragrance, and color is essential to comprehending how water and humans interact. Do not forget that the perception of colors, odors, and scents strongly affects human actions.

Water's hue has been interpreted in a variety of ways since antiquity, sparking a heated discussion that persisted until the Renaissance. Water was thought to have color much like the other three elements—air, earth, and fire—but the actual hue of water was determined by the eyes of the philosophers and scientists who investigated it or by the interior sight of the poet who addressed it. Homer refers to the ocean as a "sea of dark wine" and calls it "gray"; perhaps, this is because the ancient Greeks distinguished between colors based more on brightness than hue. But take caution—many academics remain skeptical because they contend that Homer was colorblind! Even Pindar's hyperbolic remarks on the value of water utilize color, albeit in an ambiguous way.

> Best is Water of all,
> and Gold as a flaming fire
> in the night shineth eminent
> amid lordly wealth.[22]

The river seemed white to the Athens astrologer Antiochos in the second century BC. Green and water are associated by Leon Battista Alberti in his dissertation *De pictura* (1435). Humanists of the Late Middle Ages had assigned blue or purple to water. Unlike his contemporaries, Leonardo da Vinci claims that water "takes every smell, color, and taste and by itself has nothing" in *Manuscript C*. Given that the color of water fluctuates depending on its surroundings, this statement is quite accurate. However, Leonardo never gave any details about it.

Now we know that pure water is naturally blue, though it may have additional colors due to suspended or dissolved contaminants. While comparatively tiny volumes of water appear colorless, pure water has a faint blue tint that deepens into green as the measured sample thickness rises.

The most advanced technology has been employed by modern science to answer the question of the color of water. We can now be pretty certain that water is pale blue rather than colorless.[23] With the help of a spectrophotometer, a laboratory study was conducted to observe the transient vibrations of water molecules, which ultimately resolved this conundrum.[24]

---

[22] Pindar, *Olympian Odes*, I,1, translated into English (1874) by Ernest Myers.

[23] Water owes its blueness to selective absorption in the red portion of its visible spectrum. The absorbed photons promote transitions to high overtone and combination states of the nuclear motions of the molecule, i.e. to highly excited vibrations.

[24] Braun, C. L., & Smirnov, S. N. (1993). Why water is blue? *Journal of Chemical Education*, *70*(8), 612–614.

The faint blue hue of pure water is most noticeable when observed in a lengthy water column. The blue of the water is not determined by the same light scattering that gives the sky its blue color. Rather, a portion of the visible light spectrum that is absorbed by water molecules gives water its blue color. Furthermore, to be even more accurate, the way that atoms vibrate and absorb various light wavelengths is connected to the absorption of light into water.

Because white light is selectively absorbed and scattered, the water's blue hue is an inherent characteristic. By observing a white light source across a long tube that is filled with pure water and sealed at both ends by a clear window, one can study the inherent hue of liquid water. Low absorption into the red region of the visible spectrum accounts for the pale blue hue.

On the other hand, the taste and odor of water are far more complicated to be assessed, and scientists are still debating this issue without really coming to a conclusion. While color can be reasonably explained by experimental physics, a broader range of sciences, including neuroscience, are needed to judge taste and smell. Pure water would not trigger taste but the tongue and mouth respond to water depending on compounds dissolved in saliva.

Even though Aristotle labeled water as "tasteless," it is well-known that nerve cells in the brains of insects, amphibians, and maybe mammals are sensitive to the taste of water. More recent research employing brain activity scanning techniques has demonstrated that when the tongue comes into contact with water, a part of the human cerebral cortex activates. Water may taste slightly sweet or even unpleasant since our saliva is salty.

Conversely, in odor testing, pure water is typically assessed as an odorless substance, frequently based on the olfactory capacity of a preselected group of people.[25] In practice, water molecules have no taste or smell as they are made up of one oxygen atom and two hydrogen atoms. Leonardo was right about it. However, where on Earth can we find "genuine" water in liquid form, unadulterated and pure?

Water that is completely pure and made up of just these two molecules can be created in a lab, but it is not found in nature. Since distilled water is almost completely free of contaminants like bacteria, mineral salts, and dissolved gases that are present in drinking water, it is typically purchased at the supermarket to feed certain home appliances like irons. The flavor of distilled water is less complex than that of drinking water; it is bland, tasteless, metallic, or harsh.[26]

---

[25] APAT. (2003). *Methods for measuring odour emissions.* Roma: APAT, Manuali e Linee Guida 19, based on research by Paolo Centola, Politecnico di Milano (in Italian).

[26] Shallenberger, R. S. (1993). *Taste chemistry.* Springer Science & Business Media.

Mineral salts and other substances are always present in the waters of streams, rivers, and lakes as well as in the bottled water we drink. They are the outcome of dissolved rocks of the aquifer that a water vein runs through, before coming to the surface. These compounds give the water its distinct flavor and transform the otherwise "odorless and tasteless" liquid into a resource that supports plant and animal life on Earth. Magnesium, phosphorus, and calcium are among the 500 parts per million of dissolved solids that can be found in bottled mineral water, which may also be carbonated to fit our fondness for sparkling water. However, there should not be more than 1.5 g of dissolved solids per liter in drinking water. We are able to determine that four compounds have an impact on taste perception by doing a blind tasting test on water, regardless of whether it comes from a bottle or from a tap. These are magnesium, calcium, sulfate, and baking soda.[27]

"Water takes every smell, color, and taste and of itself has nothing": apart from color, Leonardo was almost correct once more. Moreover, we enjoy the petrichor, that is, the delightful scent released when rain falls on parched ground following an extended period of drought. This is the scent that is created when a blend of bacterial materials and vegetable oils seeps into the ground during dry spells.[28] A pleasant scent, signifying the end of the rainy season and the impending return of sunlight.

The first chapter of the Leonardo's *Treatise on Water* was tentatively titled *Sull'acqua in sè* (On water itself). It was the long-coveted work for years that he was never able to finalize. Any discipline dealing with water, whether addressed to future engineers, biologists, physicists, chemists, or managers, must first recognize that water has a complicated and, in some ways, enigmatic nature. The nature of the water is a subject for much more discussion, but I must now shut up to avoid bothering the reader any further.

The Origin of Water

It takes a great deal of effort to find locations on Earth that do not reveal they are a part of a planet rich in water. Even the driest and highest deserts, such as South America's Atacama Plateau, and the largest hot deserts, such as African and Arabian ones, receive at least a few millimeters of precipitation per year on average. Although there are places where we do not yet know what the

[27] Platikanov, S., Garcia, V., Fonseca, I., Rullán, E., Devesa, R., & Tauler, R. (2013). Influence of minerals on the taste of bottled and tap water: A chemometric approach. *Water Research*, *47*(2), 693–704.

[28] Bear, I., & Thomas, R. (1964). Nature of argillaceous odour. *Nature*, *201*, 993–995.

average is because it hasn't rained for years, you can still find some water. If you bring along a portable mass spectrometer, you ought to be able to locate a few atmospheric water molecules on a desert stroll.

Ancient civilizations held different views regarding the origin of water, often attributing its existence to divine or mythological forces. Water was frequently connected to the primordial chaos that gave rise to the universe in Mesopotamian mythology. They worshipped deities like Enki, the god of freshwater, who is credited with creating rivers and springs. According to Greek mythology, the Titan god of the sea Oceanus and his wife Tethys are among the primordial deities credited with creating water. They held that all bodies of water, including rivers, springs, and seas, were connected to these divine beings. In ancient Indian texts such as the Rigveda, water is regarded as one of the basic elements of creation. It is associated with various deities, particularly Varuna, the god of the celestial ocean and the preserver of cosmic order.

Roman engineers, known for their advanced aqueduct systems and hydraulic engineering, likely had a practical, working understanding of water. They might not have had any particular ideas regarding the metaphysical origin of water. Lucretius did not inquire as to the origin of the tiny, invisible particles that compose water, despite his exploration of the natural world through both poetic and philosophical lenses.

In the western world, the dominant worldview of the Middle Ages was influenced by Christian theology, which held that God created the universe and all of its constituent parts. Water was regarded as a part of God's creation. Water was probably seen by monks to be a gift from God, necessary for life and a representation of spiritual purification in religious rituals such as baptism.

A number of natural philosophers and scientists put up cosmological hypotheses regarding the origin of water on Earth in the seventeenth and eighteenth century. Although the details of these theories differed, they all aimed to explain Earth's water supply in relation to broader cosmological processes.

The French mathematician and philosopher Descartes suggested a cosmological theory known as Cartesian Vortices to explain a number of phenomena, including the formation of Earth's oceans. He believed that the Earth began as a hot, molten mass surrounded by a swirling vortex of particles. As the Earth cooled, water vapor condensed and precipitated, eventually filling the basins of the planet and forming oceans.

In his cosmological writings, German philosopher Immanuel Kant conjectured about the genesis of celestial bodies, including Earth and its oceans.

Kant postulated that a primordial nebula, a massive cloud of gas and dust in space, provided the raw material for Earth and its oceans. Over time, gravitational forces caused the nebula to condense and coalesce into solid bodies, including planets. These planets eventually began to accumulate water through a variety of mechanisms.

The French mathematician and astronomer Laplace, who worked with Lavoisier on a number of experimental investigations, contributed to cosmological ideas of the planetary formation and the genesis of Earth's oceans. Similar to Kant, he put forth a nebular theory in his work *Exposition du système du monde* (Exposition of the System of the World). He postulated that the solar system originated from a rotating disk of gas and dust. According to Laplace Earth's oceans were a result of water vapor condensing and accumulating on the planet's surface during its formation.

Scientific knowledge of the origin of water continued to develop throughout the nineteenth and twentieth centuries as discoveries in chemistry, geology, and astronomy provided new insights into the possible mechanisms underlying the presence of water on Earth. Several key theories emerged, such as volcanic outgassing, cometary impact, solar nebula conjecture. Furthermore, the isotopic method gained popularity for analyzing the ratios of different isotopes of hydrogen and oxygen in water samples from various sources. These techniques are helpful for deducing the origins of Earth's water and its dispersal around the solar system.

The true issue is that we still do not know where all of the Earth's water came from. There are only two possibilities. According to the exogenous delivery conjecture, water was brought to Earth by icy or water-rich asteroids, after the giant impact that formed the Moon, and Earth had grown to 99% of its current size. Under the endogenous delivery conjecture, water was already inherent in the building blocks of Earth.[29]

Our idea of the formation of a rocky planet like Earth has long involved a violent, hot assembly of relatively dry material in the inner solar system 4.5 billion years ago. Therefore, water was predicted to arrive later, either delivered by rocky but still volatile meteorite infall or by comets from the frozen outer solar system. Comets, in particular, are known to contain significant amounts of water ice. The impacts of these celestial bodies could have released water vapor that eventually condensed and formed oceans.

[29] See, e.g., Fazeka, A. (2014). Mystery of Earth's water origin solved. *National Geographic*, October 30; and Gillmann, C., Golabek, G. J., Raymond, S. N. et al. (2020). Dry late accretion inferred from Venus's coupled atmosphere and internal evolution. *Nature Geoscience*, *13*, 265–269.

However, these conjectures have proved difficult to fully justify for a variety of reasons. Comets' potential contribution is limited since, for example, their deuterium concentration is often much higher than that of Earth's water. In a similar vein, isotopic variations in carbonaceous chondrites—rocky meteorites rich in water—may limit their ability to contribute to a young planet.

The solar nebula, chemical processes, degassing, and mineral hydration are a few examples of endogenic hypotheses. If there was water in the Earth's interior when it was first formed, it would have been discharged via degassing or volcanic activity. Oceans may have formed when water vapor that was expelled from the Earth's mantle during volcanic eruptions collected in the atmosphere and eventually condensed.

According to a different conjecture, water molecules were created during the accretion and formation of the Earth by chemical reactions. Under the right conditions, the elements oxygen and hydrogen, which are plentiful in the universe, may have joined to produce water.

The solar nebula theory suggests that water may have entered Earth during its formation from solar nebula material. Water-rich elements may have been part of the planet's composition as it accreted material from the protoplanetary disk.

The formation of the oceans could have been triggered by the release of water through events like volcanic activity or asteroid impacts, if water had been integrated into the Earth's rocks and minerals during their formation.

Scientists now agree that water on Earth most likely originated from a combination of events. Exogenous supply could be seen as a major source of water, such as collisions by asteroids and comets. Nonetheless, additional processes like the Earth's inner degassing and chemical reactions during planetary formation might have also been significant. All things considered, it seems plausible that a mix of these events led to the abundance of water on Earth, even though there isn't a single, conclusive explanation for the origin of water on our planet.

Rivers and water were worshipped in the riverine civilizations that arose in Mesopotamia and the Indus Valley during the third and second millennia BC, as well as in succeeding riverine civilizations and religions. Egypt differs from other river civilizations in appearance because the water and the river were not deified.

> Although the Nile was the obvious giver of life to the early men of Egypt, it was not the great river and its precious waters that first stirred thoughts of worship in their primitive minds. It was the sun, relentless bearer of death, that they supplicated.[30]

All Egyptian dynasties were strongly hooked upon the Nile. The fluctuations in the Nile flows created an inter-annual variability in the amount of available water. Still, despite its crucial and vital role in Egyptian culture, the Nile was apparently never central in their cosmology and the sun emerged as Egypt's greatest deity. From this angle, the apparent contradiction of Egyptian worship of the sun rather than that of the water can be understood because the Nile's inundation had its origin in heaven, driven by the solar disc as dictated by the myths of Horus. The water cosmology was expressed and understood through sun symbolism and the role of Hapi, the god of the annual flooding of the Nile, was minor. Not unlike some modern scientific theories, the origin of water was the sun for the ancient Egyptians.

We have so far talked and rambled a great deal about the nature of a primordial, natural, and vital fluid that is absent from nature in its purest form. Additionally, keep in mind that the perfect water for cooking ought to be low in salt and crystal clear. It should enhance the flavors of the foods without overpowering them; it should neither add nor subtract flavor. The lesson of St. Francis is that water must be humble, and those who are going to delve into its mysteries must also be humble.

[30] MacQuitty, W. (1976). *Island of Isis: Philae, temple of the Nile*. London: Macdonald and Jane's.

2

Water at Rest

For the rest, everything that concerns water is poetic
and never ceases to disturb us.
Jorge Luis Borges, La Jonction, in: Atlas, *1984.*

The Shape of Water

In a single liter of water, just sitting at room temperature, there is a mind-blowing number of molecules—about 33.54 hundred trillion! That is 3.354×10^{25} molecules, to be exact. Each one of these tiny wonders is made up of one oxygen atom, the most common element on Earth, and two hydrogen atoms, which is the most abundant element in the universe, right after helium. Imagine, a hundred thousand billion atoms in just a bit of water. That number is huge!

Over two and a half thousand years ago, Democritus, who started the Atomic School, had no clue that a simple liter of water held so many tiny particles. Neither did his followers, from Epicurus to Lucretius. They could not even dream of such a thing.

Water, like all liquids, has a special talent. It can take on the shape of whatever it is in, filling all available space. There is an old Taoist saying that goes like this:

> If you put water in a cup, it becomes the cup. Put it in a bottle, it becomes the bottle. In a teapot, it becomes the teapot.

R. Rosso, *Five Easy Pieces on Water*, https://doi.org/10.1007/978-3-031-69276-5_2

This ability of water to take on different shapes is called a "continuous medium" or continuum. We usually think of a 1-l water bottle as one whole thing. We do not think much about what's happening inside the bottle, especially when it is just sitting there. What catches our eye are its big features, like size and shape, which matter more in everyday life. The way we see things in a continuum means we are not bothered by the tiny molecular structure of the water.

In the world of classical mechanics, the idea of a continuous body or continuum is about a big, physical system. It has a low ratio between the average free path of molecules, and the size of the molecules themselves, L. This ratio is called the Knudsen number, written as $Kn = \lambda/L$, from Martin Hans Christian Knudsen, a Danish physicist known for oceanographic studies, especially on the properties of salt water. This number should not be more than 0.01 for water to be considered a continuous body. We usually think that the properties of matter, like density or viscosity, stay the same no matter where we take a sample from. This concept holds in practice until the sample size is less than one-tenth of a cubic millimiter in still water at room temperature and atmospheric pressure.

Water has this amazing quality that sets it apart from other liquids. Its particles are pretty much un-squishable, at least in most cases. If you try to squish air in a syringe and block the exit, you can shrink it by about a quarter. But try that with water, and you'll hardly see any change, even if you use all your strength (Fig. 2.1).

Pressing down on water really hard is another story. To squeeze water just a tiny bit, like one thousandth, you'd need a pressure of around 22 atmospheres. That is the same pressure you would find at the bottom of a really tall dam,

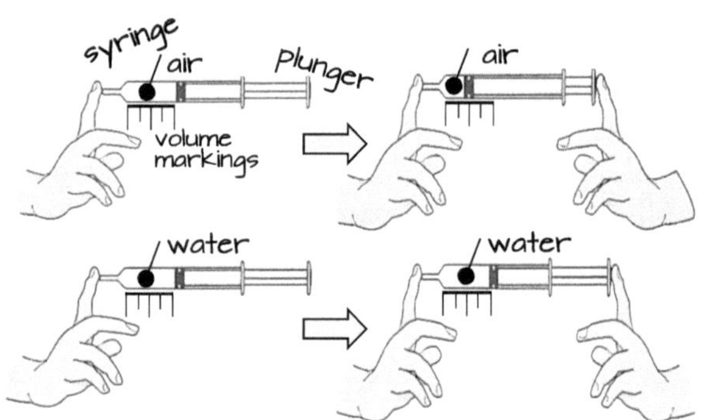

Fig. 2.1 Compression test

over 200 m high.[1] Oil needs 12 atmospheres, and mercury, a whopping 286 atmospheres. To really get how water behaves, whether it is moving or staying still, you need to understand this special trait of it being nearly un-squishable, that is incompressible. People have used this idea in all sorts of cool ways, like in the magic fountain of Heron of Alexandria from the first century BC, and the Cartesian devil, a water dancer created by René Descartes in 1640.

Buoyancy

Hydrostatics is all about studying water when it is not moving, imagining it as a continuum. Since ancient times, people have been figuring out the physical secrets of hydrostatics. Accordingly, they built stronger and faster boats, huge tanks, long aqueducts, and beautiful fountains to bring water to growing communities. The main square of Urkesh, a big archeological site from 3000 BC, was even designed to collect rainwater.[2]

About 2400 years ago, Archimedes made huge strides in this field. He lived in Siracusa, and wrote two books on hydraulics that paved the way for the systematic vision of scientific knowledge. His most famous work, *On Floating Bodies* was published around 250 BC. It is the first known treatise on hydrostatics, and Archimedes is known as the father of this field.

Archimedes' discovery is now widely known. Every kid knows that when something is put in water, it gets pushed up by a force equal to the weight of the water it pushes aside (Fig. 2.2). The origin of this idea has been told in many stories and comic books. One funny story in Vitruvius' *De Architectura* is about Hiero, the ruler of Siracusa, and a goldsmith who tried to trick him.

> Vitruvius tells us that Hiero asked Archimedes to find out if a goldsmith cheated him by mixing silver with gold in a crown. Archimedes figured it out in a bathtub when he saw water spill out as he got in. Whence, catching at the method to be adopted for the solution of the proposition, he immediately followed it up, leapt out of the vessel in joy, and, returning home naked, cried out with a loud voice that he had found that in which he was in search of. He continued exclaiming, in Greek, εὕρηκα, Eureka! (I have found it out). He realized that he could use this idea to test the crown. He compared how much water spilled over

[1] Temperature affects the compression ratio. In water, this ratio decreases by about a fifth when you go from five to eighty degrees Celsius. Liquid water retains its high incompressibility even at high temperatures.

[2] Buccellati, G., Ermidoro, S., & Mahmoud, Y. (2018). *The millennia for today. Archaeology against War: Yesterday's Urkesh in Today's Syria*. Malibu: Undena Publications.

Fig. 2.2 Archimedes' experiments

> when he put in gold, silver, and the crown. This way, he found out the crown wasn't pure gold.

The force that pushes up an object in water is called Archimedes' force. It shows how important experiments are in understanding water. Leonardo da Vinci also believed in experimenting, not just theorizing.

> Remember to start with the experience and then move on to the reason when arguing about waters.[3]

We call Archimedes' force hydrostatic buoyancy, but isn't just about water. It works in any liquid or gas. It is pretty simple to calculate. The force F equals the density of the water ρ, times gravity g, times the volume V of the displaced water. This law, $F = \rho g V$, helps us understand why things float or sink.

The volume of displaced water V is the volume that the floating body is immersed in, which is measurable in cubic meters, m^3. Gravitational acceleration, g, is approximately 9.81 meters per square second, m/s^2. The density of water, that is the ratio of mass to volume is about 999 kg/m^3, for freshwater at 15° C. If a boat is sailing with a submerged volume of 10 m^3, the buoyancy force equals approximately 98 thousand Newtons. The Newton N is the unit for measuring force, the amount of force needed to give a kilogram of mass an acceleration of 1 m/s^2. According to the International System of Units, N denotes the force giving a mass of 1 kg an acceleration of 1 m/s^2, $1\ N = 1\ kg\ m/s^2$.

Archimedes' law ensures that the volume fraction of an immersed object is equal to the ratio of body and liquid densities, that is $\rho_{object}/\rho_{water} \leq 1$ (Fig. 2.3). Since a floating iceberg has a density of about 917, while the density of salt water is about 1025, its immersed volume is roughly 90% of that of the ice's block.

Take the Titanic, for example. It was a huge ship, the biggest and fanciest of its time. But it hit an iceberg and sank, killing about fifteen hundred people. A few days before, Captain Edward Smith had said that "the big icebergs that drift into warmer water melt much more rapidly under water than on the surface, and sometimes a sharp, low reef extending two or three hundred feet beneath the sea is formed. If a vessel should run into one of these reefs half of her bottom could be torn away." He didn't know that his ship would suffer this fate. The Titanic was huge, but unlike the iceberg that sank it, most of it was above water. Despite its gross tonnage of about 46 thousand tons and a

[3] Leonardo da Vinci, *Manuscript H*, Folio 90R, T1: "Ricordatj quando comentj l'acque dal legar prima la sperienza e poj la ragione".

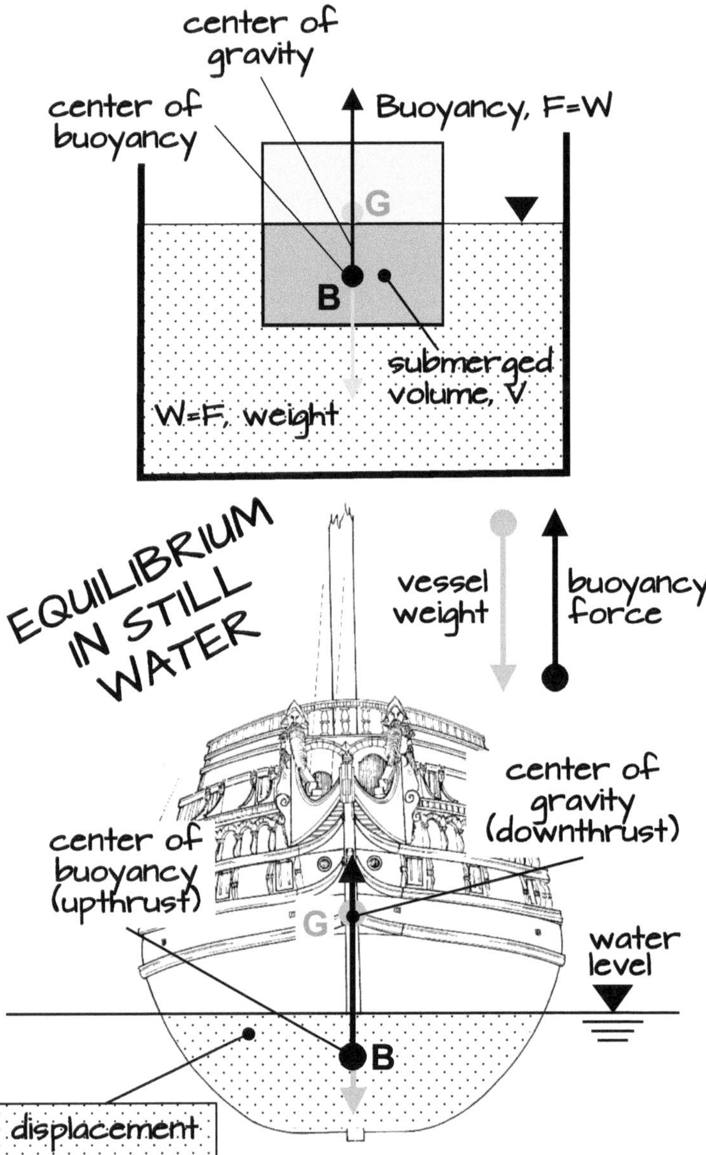

Fig. 2.3 Buoyancy

displacement of 52, the draught of the Titanic was only 10 and a half meters. Unlike the iceberg that sank it, the large ship stood more than forty meters above sea level. Nowadays, a 200,000-ton sea monster that sails with a 16-m draught emerges from the water like a 10-story building.

Archimedes' law is really about Newton's second law of dynamics. It says that the force on a body depends on its weight, which comes from its density and volume, ρV. To float, a ship must balance the force that lets it float.

To remain afloat, it is essential to know the force that allows an object to float. Take the Italian aircraft carrier Cavour, for example. It needs a certain amount of water displaced to float. It is about 244 m and weighs more than 27,000 tons when it is fully loaded. This weight is equal to the amount of water the ship pushes out of its way when it is in the water. The part of the ship under water takes up about 27,000 m^3 of space. To figure out how deep the ship goes into the water, you subtract this space from where the ship's bottom meets the waterline. Its stability depends not just on the buoyancy force, but also on where this force is applied. Also, the depth of the submerged part of the hull allows to assess, for example, the capacity or otherwise of a port to accommodate the vessel. Cavour's draft is 7.5 m.

Equilibrium

The Vasa, a huge warship, was 69 m long and nearly 12 m wide. This galleon weighed 1270 tons and could float just fine, yet it sank on August 10, 1628, only 120 m from the shore, right after it was launched.

Why?

An object remains buoyant if its buoyancy force equals its weight. Its stability, however, is more complex stuff. In the stability of floating bodies, the stable equilibrium is attained if the metacenter (M) point lies above the center of gravity (G). The metacenter—the point about which a body starts oscillating when the body is tilted by a small angle—is the point where the line of action of buoyancy will meet the normal axis of the body when the body is given a small angular displacement (Fig. 2.4). When the metacenter and the center of gravity coincide at the same point, then the body is in neutral equilibrium. Otherwise, the equilibrium will be stable or unstable depending on the location of the metacenter. If it is above the center of gravity of the object, then the disturbing couple is balanced by restoring couple, thus the body will be in stable equilibrium. Conversely, if the metacenter is below the center of gravity, then the disturbing couple is supported by restoring couple, and the equilibrium of the object will be unstable.

Stability is compromised when a ship's center of gravity is higher than the metacenter, leading to a greater risk of overturning. This principle played a crucial role in the fate of the Vasa, altered by the Swedish king's decision to add a second gun deck. This addition, while transforming it into a formidable

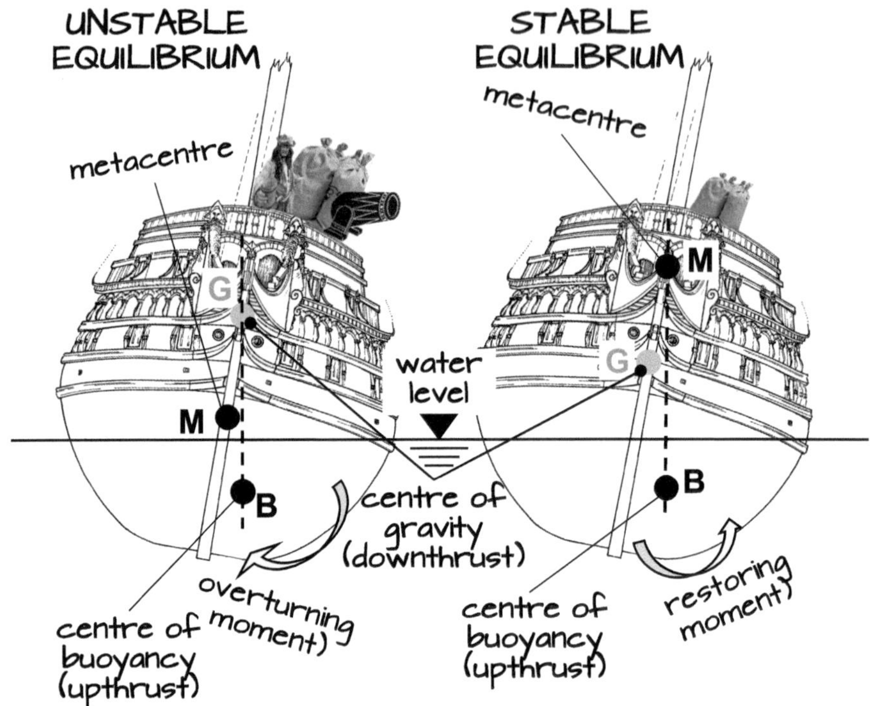

Fig. 2.4 Equilibrium stability of a floating body

warship of its era, significantly raised the vessel's center of gravity above the buoyant center. The helmsman, displaying remarkable skill, managed to right the ship during the first wind gust. Unfortunately, a second gust caused the ship to lean, taking on water in front of the gathered Stockholm populace, who were there to celebrate its maiden voyage. Tragically, the Vasa sank rapidly, having sailed less than a mile.

The boat's displacement depends on its dive, which varies between empty and loaded states. The sinking of the Vasa was caused by insufficient initial stability, as the ship couldn't resist skidding—a phenomenon triggered by wind and waves. The mass distribution within the hull—including ballast, supplies, and other loaded items—was too high above sea level, extending beyond the second deck of guns. This elevated the center of gravity significantly above the center of buoyancy, making the ship prone to tilting with minimal force and lacking sufficient righting moment to return upright. Even in calm seas and light winds, a ship with this design flaw could experience severe slipping.

In the case of the submerged body the center of gravity and center of buoyancy is fixed, therefore the stability or instability is decided by the relative positions of the center of buoyancy and the center of gravity. The body's center of gravity must lie below the center of buoyancy of the displaced liquid for a submerged object to be stable because the disturbing couple is countered by the restoring couple (Fig. 2.5). If the center of gravity is above the center of buoyancy, the body will be unstable. A submerged object is in neutral equilibrium for all positions if the body's centers of gravity and buoyancy overlap.

Conjectures and Experiments

Archimedes' discoveries faced greater challenges than one might expect, a fact highlighted by a lesser-known incident in Galileo's scientific career. In 1611, a group of Tuscan Aristotelians questioned Galileo's credibility following a dispute about why bodies float in water. This controversy was fueled by the Tuscan academy's eagerness for the honor of being named Primary Mathematician and Philosopher to Grand Duke Cosimo II. The argument escalated after Galileo countered Aristotle's belief that ice is condensed water,

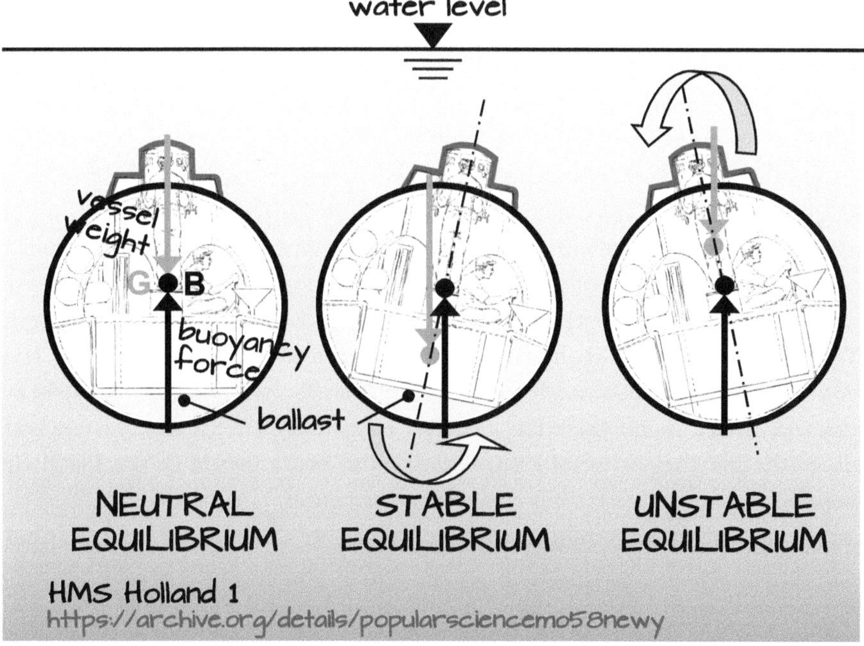

Fig. 2.5 Equilibrium stability of a submerged body

asserting instead that ice is rarefied water, lighter and therefore buoyant in liquid water.

In discussing the differences between condensation and rarefaction, the issue of buoyancy quickly became central. This involved determining what makes an object either float on a liquid's surface or submerge. Aristotelians argued that an object's shape, like that of ice, dictates whether it sinks or floats. They believed the varying shapes of objects determine their behavior in water. Contrarily, Galileo supported Archimedes' hydrostatic principle, stating that buoyancy depends on the difference in specific weight between the liquid and the submerged solid.

Galileo, responding to criticism, created a balance that transformed conjectures into experiments. This instrument measured the weight of objects in both air and water. Earlier, Archimedes had developed a buoyancy-based method to determine that a crown was of silver-gold alloy, not pure gold. Unlike the simpler solution described by Vitruvius, Galileo sought a more complex approach, also relying on buoyancy but with strict hydrostatic principles. His technique involved a dual weighing system, one in air and the other in water (Fig. 2.6). Although innovative, Galileo's balance wasn't entirely original. It had predecessors in the water or hydrostatic balance, the immersion hydrometer, the pycnometer, and the triple hydrostatic balance.[4]

Galileo ingeniously combined the behavior of floating sticks with that of hollow bodies in his solution. He proposed that air not only supports but also lifts and propels these objects upward. This sophisticated theory allowed him to explore the structure of water and other fluids through experimentation rather than mere speculation. Despite claiming some uncertainty, Galileo posited that liquids consist of closely packed particles or atoms, held together merely by contact. This concept explained fluids' high penetrability, suggesting it does not obstruct but merely slows down an object moving through it.

To grasp the concept of still water, one must train the human eye to discern various scales, both spatially and temporally. Leonardo da Vinci, in his *Manuscript A*, states that the core of the Earth element lies equidistant from the ocean's surface, but not from the Earth's surface. He notes, "for it is evident that the globe nowhere has a perfect roundness, except where there is the sea, or the marshes or other calm waters. And every region of the Earth that rises above the water is farther away from the center." Leonardo theorized that water's erosive power would ultimately shape the Earth into a perfect sphere.

[4] Mottana, A. (2017). *Galileo e la bilancetta. Un momento fondamentale nella storia dell'idrostatica e del peso specifico*, Biblioteca di «Galileana», vol. 7, Firenze: Casa editrice Leo S. Olschki (*Galileo and the balance. A fundamental moment in the history of hydrostatics and specific gravity*, in Italian).

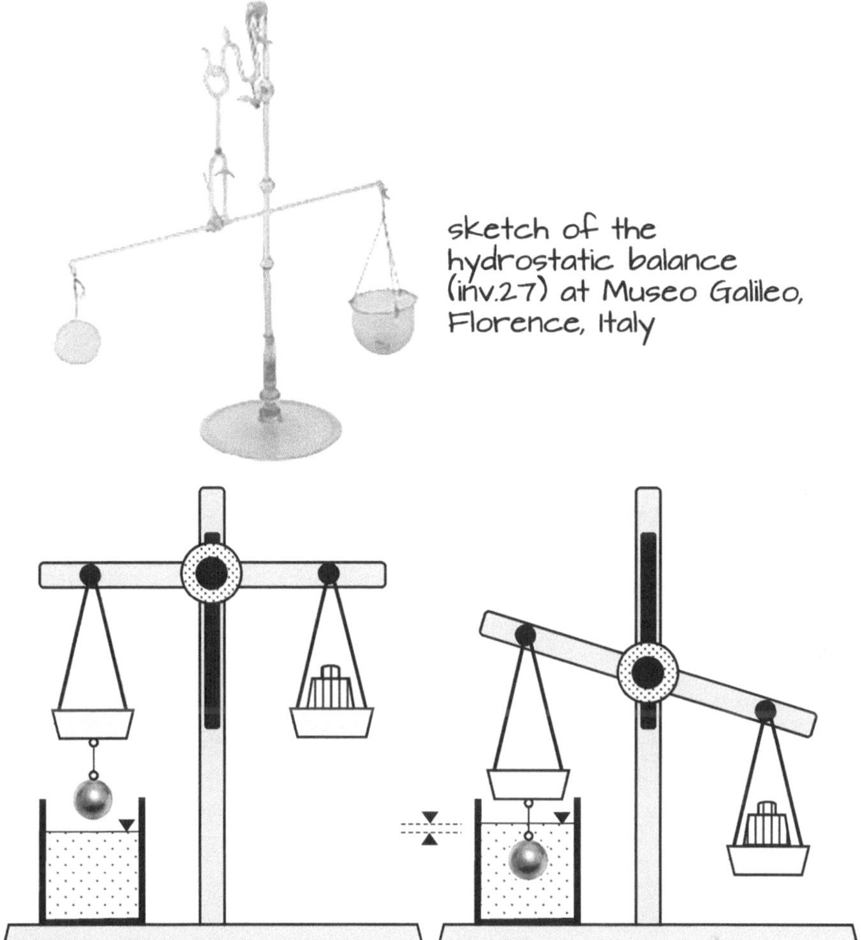

Fig. 2.6 Hydrostatic balance

He initially imagined a vast cave filled with water at the Earth's center, surrounded by water. However, in the *Codex Atlanticus*, he revised this notion, noting that a significant portion of the cave had collapsed toward the Earth's center, attributing this to the veins that continuously erode the areas through which they flow.

While Leonardo's perspective might seem fanciful, it embodies a prophetic vision when merged with the concept of buoyancy. Archimedes' principle, in fact, applies globally in explaining isostasy—the floating of the Earth's rigid crust on its more fluid mantle. Isostasy results from an increased thickness in the crust's rock during orogenic processes and from the growth of ice sheets at high latitudes during ice ages. In both scenarios, the asthenosphere (the mantle's upper layer) sees rocks sinking due to added weight. This sunken region, or "root," experiences a series of isostatic adjustments to maintain equilibrium with the asthenosphere, facilitated by its plastic behavior (Fig. 2.7).

Leonardo da Vinci consistently revisited his cosmological puzzles, seeking a rationale for the transition from macro to micro scales. In the *Leicester Codex*, he noted that water has three focal points: a global, planetary center, a local center, and the center of individual water droplets. He described the global center as encompassing all stationary waters—including canals, ditches, ponds, fountains, wells, stagnant rivers, lakes, and seas. Despite their varied elevations, he observed that each body of water's surface is equally distant from the Earth's center. This is evident in lakes situated atop lofty mountains. The spherical form of a water droplet, he concluded, mirrors the shape of this universal element.

Leonardo lacked the means to experimentally test his cosmological insights. In contrast, Archimedes transformed his intuition into a practical tool, the hydrostatic scale. This invention not only sparked curiosity but also, thousands of years later, inspired the inventiveness of Galileo.

The Archimedes scale measures the upward force experienced by a solid when submerged in a liquid. This device features two plates; one is shorter and has a hook for attaching various accessories, typically brass cylinders of different weights. These cylinders are submerged in a water pan. Upon immersing the solid cylinder in water, the scale becomes unbalanced. To restore balance, the hollow cylinder is filled with water until the solid cylinder is fully submerged. Equilibrium is achieved when the Archimedes force matches the weight of the displaced water volume.

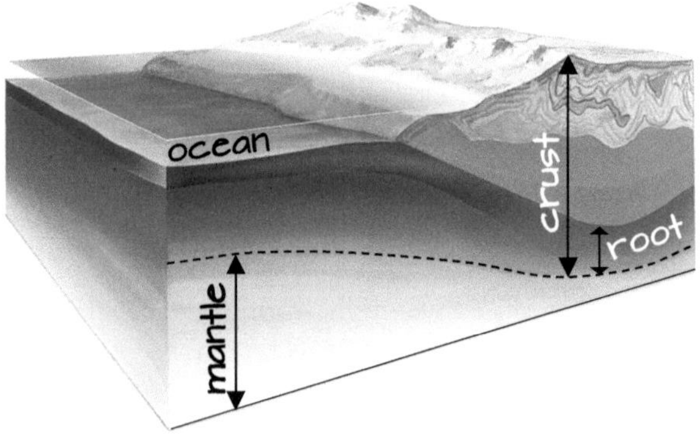

Fig. 2.7 Isostasy

Beware though! The technique of determining an object's weight by gauging the volume of displaced water works well for floating bodies. Yet, it falls short when the object is fully submerged. This is because the rise in water levels reflects the object's volume, not its mass. The exception is when the object's density precisely matches that of the fluid.

Pressure

Leonardo's miter gate, an exemplary original at the Gabelle Lock in downtown Milan, Italy, expertly manipulates hydrostatic forces to control the water levels of the Martesana Canal. This mechanism consists of two fixed-height, movable barriers, sealed securely by the water load arising from varying upstream and downstream water levels. The design resembles a hinged double door, with each door attached to the shoulder wall.

The two doors close against each other at an internal angle smaller than 180°, typically around 120°, forming a shape similar to an arrowhead (see Fig. 2.8). This mechanism operates like a swing-type check valve, functioning as a non-return system and operating under pressure levels comparable to those in a bicycle's inner tube.

Leonardo's design for his miter gate was guided not by calculations but by intuition and practice. It wasn't until Stevin's law in 1586 that the ability to accurately assess water pressure on the doors became possible.[5] This pressure, the force per unit surface from the weight of a stationary liquid, is directly proportional to water density and depth. This proportionality hinges on the acceleration of gravity. Pressure is determined by the product of three factors: $p = \rho g h$. Here, p represents pressure, measured in Newton per square meter and h corresponds to hydraulic head, which indicates the water depth in meters. Upstream from the gate, hydrostatic pressure is directly related to the water height within the channel. Conversely, the downstream pressure is proportional to the lower water head. Consequently, the gate's closing force depends on the elevation difference between the upstream and downstream water basins.

Many historians question whether Leonardo da Vinci truly pioneered this technique, which has historical precedents stretching back a millennium in the Far East. Indeed, Giovanni Battista Alberti anticipated it in his work *De Re Aedificatoria*, which was conceived using the Vitruvian model of *De*

[5] Stevin, S. (1586). *De Beghinselen des Waterwichts*. Leyden: Druckerye van Christoffel Plantijn (*Statics and hydrostatics*, in German).

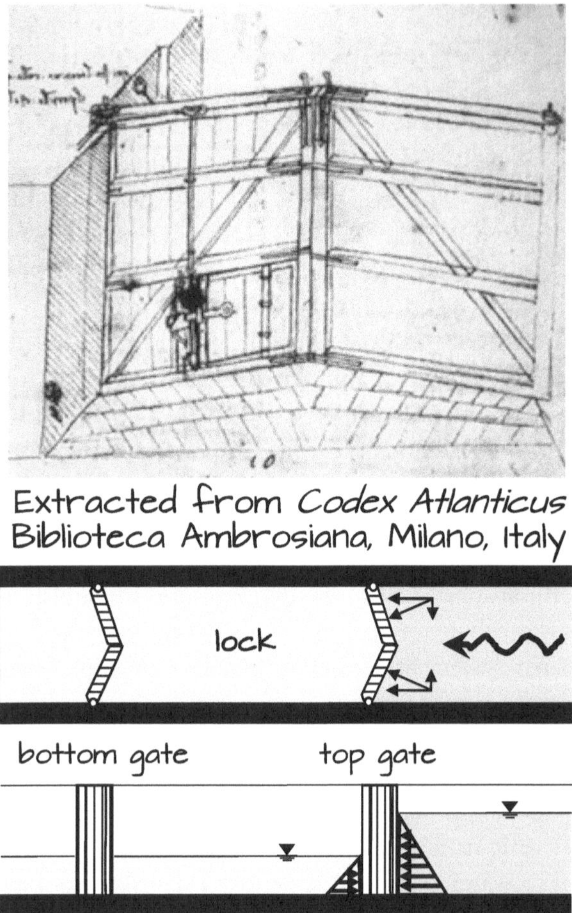

Fig. 2.8 Da Vinci's lock

Architectura and published in Florence in 1485. Nonetheless, the *Da Vinci Mitre Lock* is the common term for this system, which controls navigation locks worldwide. When passing through the Panama Canal between the Pacific and Atlantic seas, you can still see the gigantic gates of the da Vinci's locks control, which manage the flow of ships across the basins that connect the two.

Archimedes, Vitruvius, and Leonardo are well-known for their engineering and architectural skills, and Simon Stevin, also known as Simon of Bruges in his homeland, was not far behind. His position as Intendant General of the Public Works of the United Provinces required him to construct dams, canals, water mills, and even a sailboat.

Water at rest is an important structural component in many civil and military buildings because it contributes significantly to static stability. A force of approximately one thousand pounds burdens a one-meter-wide vertical wall that is wet on one side by a two-meter-high water basin. This load is significant because it can either stabilize or unbalance the structure depending on the scenario. The creation of ditches filled with a substantial volume of water safeguarded old castles. The hydrostatic load against the wall had a stabilizing effect. It could be a terrible idea to drain the pit because it might result in catastrophic failures (Fig. 2.9).

Stevin's law is also derived directly from Newtonian dynamics. The pressure p operating at any particular position is the force applied per unit of surface to the submerged body. The equation is equal to the product of gravity's acceleration g and the specific mass of the water above it, calculated as the product of density ρ and depth or "head" h. The Pascal unit, named after Blaise Pascal, a seventeenth-century French scientist, mathematician, philosopher is used to measure pressure. Shortly, we will find out why. One Pascal equals one Newton per square meter.

Liquid water is an excellent example of an incompressible fluid. When the pressure of a restricted fluid increases, it spreads throughout the container. The intensity of the increase remains constant, but it moves in the same direction, always perpendicular to the container's wall where the fluid exerts pressure. Pascal's law (1653) is an important theoretical discovery, but it has also proven useful in a wide range of practical applications. The use of this law

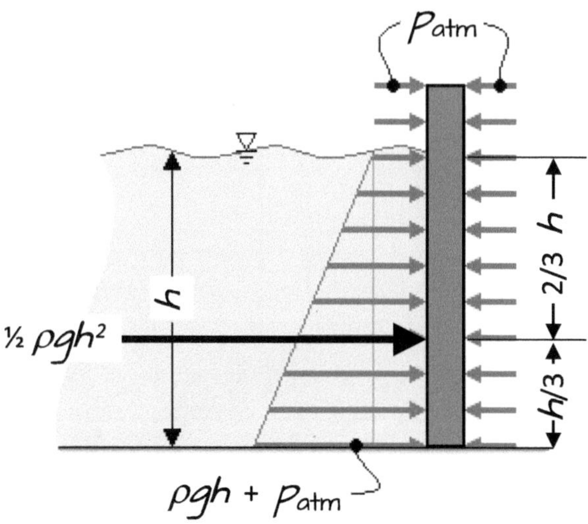

Fig. 2.9 Pressure distribution

enabled the building of the siphon and hydraulic press, the hydraulic press and jacks capable of lifting massive weights, the aqueduct tower tanks, and the hydraulic brakes that equip modern automobiles.

Pascal's law is derived from Stevin's law and always applies under the hypothesis of an incompressible fluid. This law explains how pressure changes. The pressure differential is calculated using three multiplicative factors: water density, gravity's acceleration, and hydraulic head loss (the difference in water level between upstream and downstream): $\Delta p = \rho\, g\, \Delta h$. Over-pressure Δp is measured in Pascal units, whereas head loss Δh is measured in meters (m).

To demonstrate, Pascal put a 10-meter-long vertical pipe into the top of a water-filled barrel. He filled the barrel to capacity by putting additional water into the pipe. The barrel's pressure grew as the water level in the pipe rose, forcing it to explode (Fig. 2.10). This experiment discusses the hydrostatic paradox, sometimes known as Pascal's Paradox. It is a hydrostatic

Fig. 2.10 Pascal's experiment

Fig. 2.11 Pressure head

phenomenon in which pressure at a resting point in water is unaffected by the shape of the container and only depends on the depth of the point below the water's surface. This means that the pressure at a particular depth remains constant in all directions, regardless of the container's shape.

To confirm this, one can measure the pressure at the bottom of two distinct vessels, such as a narrow glass bottle and a potbelly glass jug with the same bearing surface and water level. The gauge will show the same pressure, but the jug's weight will be substantially higher than the bottle's, even if the two empty vessels weigh the same (Fig. 2.11).

Simple experiments permitted humanity to achieve enormous advances. Leonardo da Vinci's manuscripts contain numerous pictures that show how hydrostatics can be treated through experimentation rather than theoretical speculation. Numerous patterns of connected pipes filled with liquids of different densities are recorded in the *Leicester Codex*, as are numerous sketches of orifices and hydraulic weirs, and multiple efforts at measuring pressure. At the time, practically everyone, including Leonardo, dismissed the concept of pressure since they were unfamiliar with Archimede's treatise *De insidentibus aquae*, despite the fact that Willem van Moerbeke translated Archimedes' book, which introduced this concept, into Latin in the thirteenth century.[6]

[6] Arredi, F. (1942). Le origini dell'idrostatica. *L'Acqua*, nos. 2, 3, 7, 11, 12 (The origin of hydrostatics, in Italian).

The Cartesian diver is a classic scientific toy and experiment that illustrates Archimedes' and Pascal's ideas. It is often made up of a small container with a sealed lid (such as an eyedropper or a small glass ampoule, sometimes fashioned like a devil) that is partially filled with air and immersed in a larger container of water. The diver is half filled with water and has a pocket of air trapped inside. When the larger container is squeezed, the pressure within rises. This pressure is conveyed through the water, compressing the air within the diver. As the air compresses, the diver absorbs more water, which increases its density and causes it to sink. When the pressure is lifted, the air expands, decreasing the diver's density which allows it to float back to the surface. This experiment exhibits multiple scientific ideas, including Pascal's principle of fluid pressure transmission, Archimedes' principle of buoyancy, and Boyle's law, which explains the inverse relationship between gas volume and pressure.

Fish use a similar method to travel vertically. The swim bladder, also known as the air bladder, is an internal gas-filled organ that allows many fish (excluding cartilaginous fish) to adjust their buoyancy and hence remain at their current water depth without expending energy by swimming.

This is how submarines function. The waterproof ballast tanks, which may be filled or emptied, let it float, sink, or halt at a certain depth. When the tanks are full, the submarine's weight grows and the buoyancy can no longer support it, causing the submarine to dive. Emptying the ballast tanks, on the other hand, reduces the submarine's weight and causes it to rise to the surface.

Gravity and Capillarity

Pellegrino Artusi, the pioneer of kitchen science,[7] described cooking as "a tricky nymph." Water is also an odd nymph. It escapes on its own and needs to be contained. However, depending on the design of the container, it may move even when at rest. It is immobile, but it moves. And he alludes to his ability to maintain balance much like a bicycle sprinter on the track.

Heron of Alexandria, who lived twenty-first centuries ago, created the water-powered organ, the *Aeliopile*, a rocket-like reaction engine, the first steam engine, and a number of water machines. The *Heron's Fountain* is a free-standing fountain powered by self-contained hydrostatic energy. It is a unique mechanism that is designed so that the jet's height exceeds the elevation of the water surface in the basin, with no concealed pumping stations. This action appears to contravene fundamental physics rules, particularly those that

[7] Artusi, P. (1891). *Op. Cit.*

govern hydrostatic pressure. Centuries before Archimedes and Erone, Aristotle offered the idea that water seeks its own level, resulting in the concept of hydrostatic equilibrium. Perhaps Heron's bargain was to insult the most prominent philosopher of all time?

The device was made up of three containers at different elevations: one on top (A), another at the bottom (C), and an intermediate one (B). In contrast to the atmospheric pressure open top tank, both the intermediate and bottom tanks are airtight. The three tanks were connected by three pipes. The first (P1) joins tank A and C via a hole in the bottom of the basin (A) and the bottom of the air supply container (C). The second (P2) connects the bottom of tank C and intermediate B, running from the top of the air supply container (C) to the top of the water supply container (B). The third (P3) connects tank B to basin A, running from the bottom of the water supply container B to a height above the basin's rim. The fountain emerges from this pipe. The maximum height of P3 pipe is determined by the height between B and C. To start the fountain, the upper tank must be full of water and the lower tanks empty.

By opening the outlet of pipe P1, which joins the upper and lower tanks, the air inside is compressed and forced away through pipe P2, which connects tank C to intermediate tank B, forming an inverted siphon. Air exerts pressure on the water in the intermediate vessel, which flows upward through pipe P3 with a nozzle before exiting at atmospheric pressure. The water that comes out is enthusiastically poured, flowing into the higher cup as if from a sprayed jet (Fig. 2.12).

The energy for moving the water is ultimately derived from the water in tank A descending into C. This indicates that the water in tank B can only rise into A to the same extent that it descends from tank A to C. Water dropping from A to tank C via pipe P1 creates pressure in the bottom container, which is proportional to the height difference between tanks A and C. The air transmits pressure through pipe P2 into the water supply B, pushing the water upward into pipe P3. Water flowing up pipe P3 replaces water falling from tank A into C, completing the loop.

Heron's fountain is not a perpetual motion mechanism. If the nozzle of the spout is narrow, it may run for a few minutes before stopping. Unfortunately, the operation will only end when the interim tank is empty. The water coming out of the tube may rise above the level of any container, but the overall flow of water is downward. However, if the quantities of the air supply and fountain supply containers are built to be much larger than the volume of the basin, and the flow rate of water from the spout's nozzle is kept constant, the fountain should be able to function for a much longer period of time.

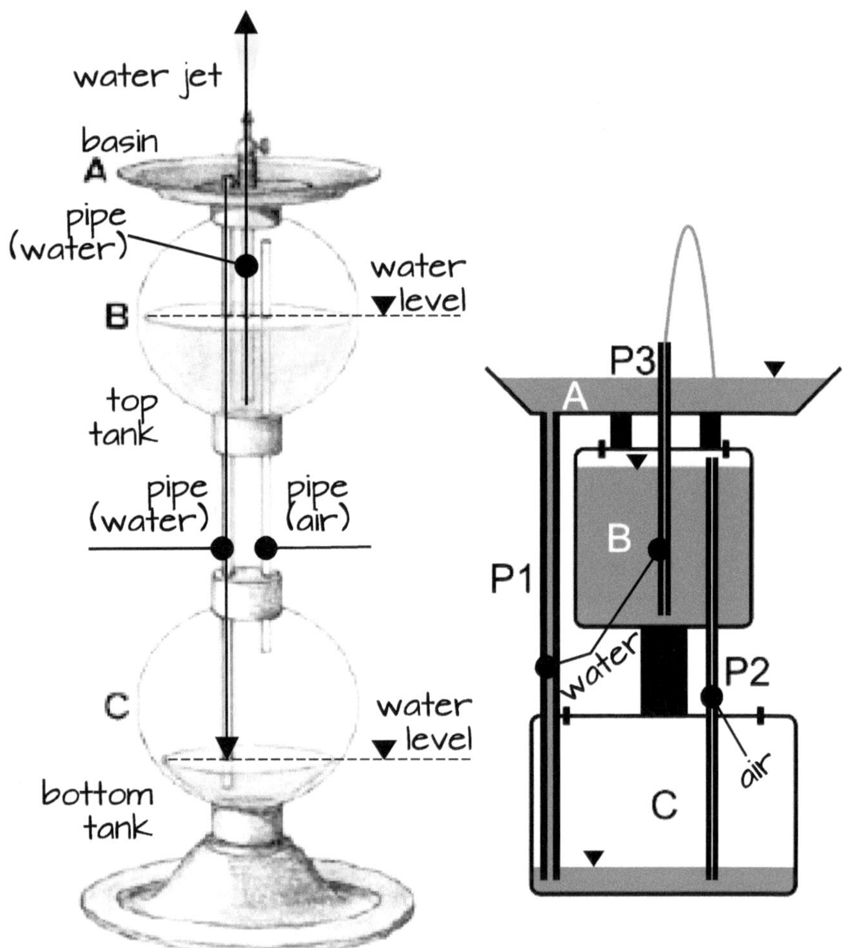

Fig. 2.12 Heron's fountain

Hydraulic scholars commonly face the problem of continually moving hydraulic machines, which are frequently used by more or less eccentric innovators. Because of their complexity, the refutation can be difficult. The obsession of perpetual motion has motivated renowned scientists such as Leonardo da Vinci. A remarkable study of a perpetual motion pump can be found on a Leonardo sheet dedicated to water-lifting systems. A notable element is the feeding device, which employs a hydraulic wheel with three compartments that are fed via a pair of pumping pistons powered by cams positioned on the wheel's axis. Proving that perpetual motion cannot occur in this brilliant machine is a difficult task for any scholar (Fig. 2.13).

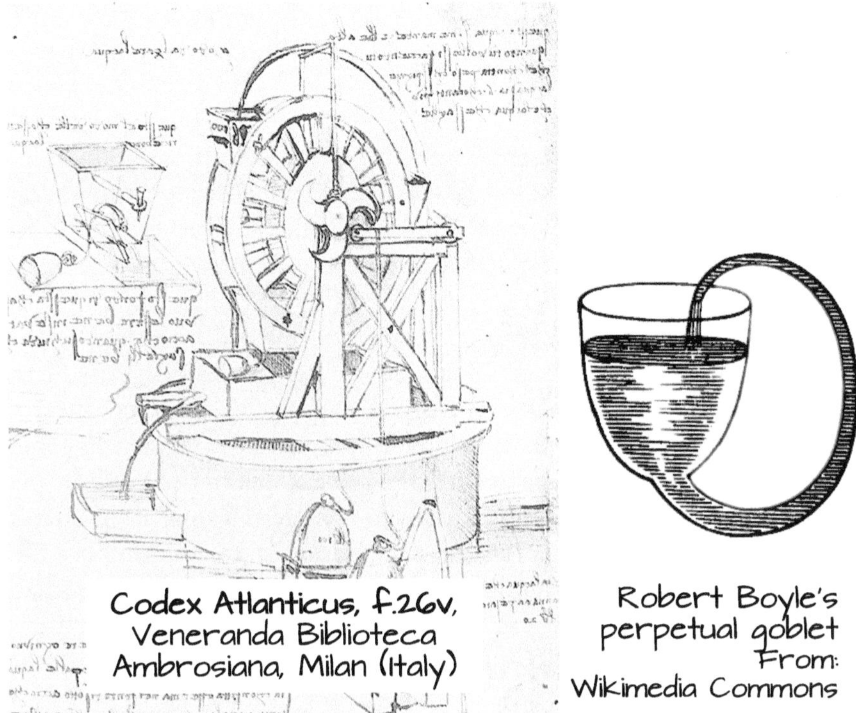

Fig. 2.13 Perpetual motion

Leonardo is not alone: the history of perpetual motion machines dates back to the Middle Ages. The possibility of perpetual motion machines remained uncertain for millennia since modern thermodynamics theory only has demonstrated that they are not feasible. Water played a big role in this compulsive search, from the water screw by Paracelsian physician Robert Fludd to the bucket fountain by Edward Somerset, the man who is thought to have made the first steam engine.[8] The first modern chemist, Robert Boyle, spent some time on his self-flowing flask, a perpetual motion machine. He called it "hydrostatic perpetual motion" because it appears to fill itself through siphon action. It is simply a hydrostatic paradox, impossible in reality. A siphon requires its output to be lower than the input.

The action of Heron's fountain may appear less paradoxical if viewed as a siphon, with the upper arch of the tube removed and the air pressure between the two bottom containers providing positive pressure to lift the water over the arch. The gadget is also known as an inverted siphon. The gravitational

[8] Ord-Hume, W. J. G. (1977). *Perpetual motion: The History of an obsession*. New York: St. Martin's Press.

potential energy of the water that falls from the basin into the bottom tank is transmitted via a pneumatic pressure tube to push the water from the upper container a short distance above the basin. At this point, only air is being propelled upward.

Gravity drives many macroscopic water processes. Gravity drives the change of stationary underground water into a jet that gushes from the earth, as it does in artesian wells, where groundwater rises to the surface without the need of artificial aids. Water rises to the hydrostatic, or piezometric, level, where the water pressure in the aquifer is zero, equal to atmospheric pressure.

An artesian aquifer is a layer of permeable soil, such as limestone or sandstone, sandwiched between two impervious or less permeable layers, such as clay or rock (Fig. 2.14). The top level of the water table determines the aquifer's pressure if the strata are concave and have an inverted saddle comparable to that of a basin. This causes the water level in the well to rise until the pressure equals the weight of the water, putting it under pressure. If the natural pressure is high enough, water can reach the earth surface, resulting in a flowing artesian well. Water returns to aquifers when the water table in the recharge zone is higher than the well's head. This technology has been known in Europe since around 1000, when Carthusian monks in Artois first used it to water their fields.

Archimedes, Vitruvius, Stevin, and Pascal concentrated on the most evident macroscopic behavior of water at rest, which is mostly determined by gravity. Water can be subjected to forces other than gravity even when at rest, depending on the container. This force can have a significant impact on water behavior in some conditions.

When water is kept in a pot with a small section in comparison with its length and width, the influence of surface tension becomes noticeable and

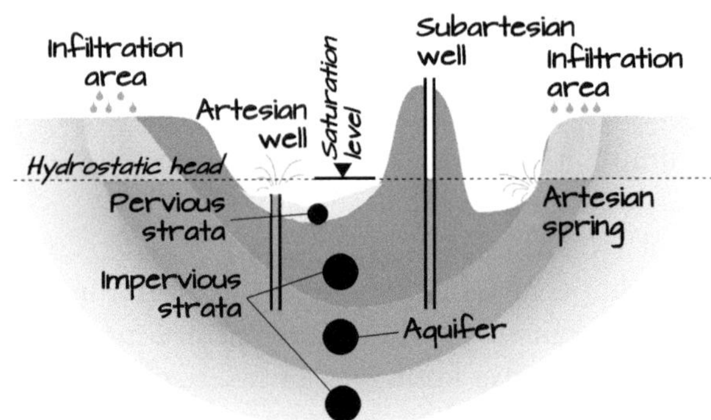

Fig. 2.14 Groundwater head

gradually dominates. Capillary action, which lifts water by sticking to the mouth walls, causes a curve in the free surface comparable to that of the knee bones. For this reason, the skin's surface is known as a meniscus, and it is a common occurrence when we drink through a straw (Fig. 2.15).

When water's free surface interacts with a gas, such as the atmosphere or vacuum, it is clear that something is happening. Because water cannot endure tensile and shear stress, the free surfaces immediately settle into an equilibrium state. At a small scale, the equilibrium of two materials, such as water and air, is controlled not only by gravity, but also by surface tension. At this scale, intermolecular forces exist between the liquid and the surrounding solid surfaces.

Capillary action is the mechanism by which water flows in a tight space without the assistance of, or even in opposition to, other forces such as gravity. It can be seen in the accumulation of paint between the hairs of a paintbrush, in porous materials like blotting paper, in loose materials like sand, in the wick of a candle, and in a narrow tube like a straw. If the straw's diameter is small enough, the combination of surface tension induced by cohesiveness within the water and adhesive forces between the water and the container wall will propel the liquid. Similarly, capillarity can carry water in the other direction, as in dry soil that has recently been irrigated by sprinklers. The soil's surface will dry quickly as a result of rapid percolation along the pores, until water reaches the ground table, where pressure equals atmospheric pressure. If you maintain watering, the process will continue until all the soil is saturated.

Humidity descends from below in loamy or clayey soils. The power that pulls it is comparable to that exerted by the wick of an oil lamp or candle. We

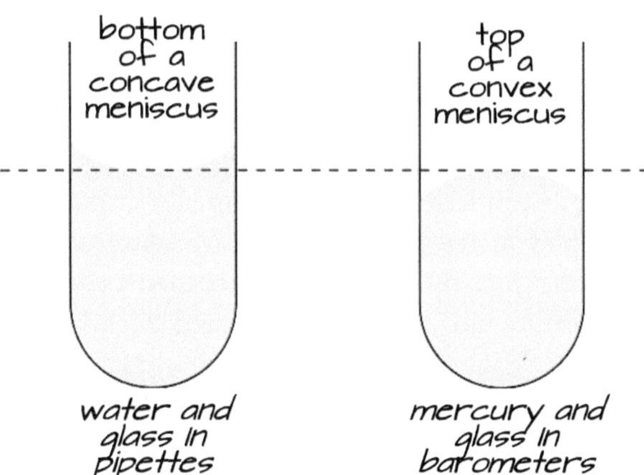

Fig. 2.15 Capillary interface

see it by immersing one end of a length of dry string in a water pan and hanging the other end over the edge, with the tip at a level below the pan's water level. The water will gradually climb the string, saturating it throughout its length until it drops from the top. A drop of water usually leaves a thin film on small particles like string twine or dirt particles.

Leonardo da Vinci recorded numerous observations in the *Codex Leicester.*[9] For instance, next to the splash of a rag overflowing from a pail, Leonardo wrote:

> As water passes through the felt, the weight of water in holes of the vessel is offset by the weight of water on the folded felt inside the vessel.

These statements accurately describe capillarity physics, which was later investigated by scientists such as Robert Boyle, Jacob Bernoulli, and Giovanni Alfonso Borrelli in the seventeenth century. French Pierre-Simon Laplace and the English Thomas Young were the first to formalize knowledge mathematically in the early nineteenth century. Understanding capillary has always piqued the interest of renowned scientists, including Gauss and Kelvin. Capillarity was the subject of Albert Einstein's first paper, published by *Annalen der Physik* in 1900.

Capillary action causes porous medium to get damp. This process, while not immediately visible, is important for providing plants with the water they need to survive and thrive. Soil moisture is thus kept even during dry seasons when there is no rain and the groundwater level is low, far from the roots. The capillary effect also has a significant impact on biological systems, facilitating water flow in the plant's xylem.

The capillary rise height is inversely proportional to the opening's diameter. Water rises a few centimeters in coarse-grained soils like sand, but it can rise several meters in very fine-grained soils like clay (Fig. 2.16). In addition, the degree of saturation, or the percentage of soil voids occupied by capillary water, changes vertically and decreases from the bottom up. It drops from the top down when water infiltrates the soil.

Despite being a wonderful legacy, Archimedes' ceramic bathtub, which is presently on display at the Archaeological Museum of Syracuse, asks some awkward questions: had the scientist ever bathed before? If he washed frequently, why didn't he notice that water poured out when he entered the full bathtub? Lawrence Wright believes that bathrooms can teach us more about a community's history than battles. During severe droughts, when the earth lay parched under a relentless sun in summer 2022, an environmentalist kindly

[9] See. e.g., *Codex Leicester*. Folio 34V 44–49 and Folio 27R 1–13.

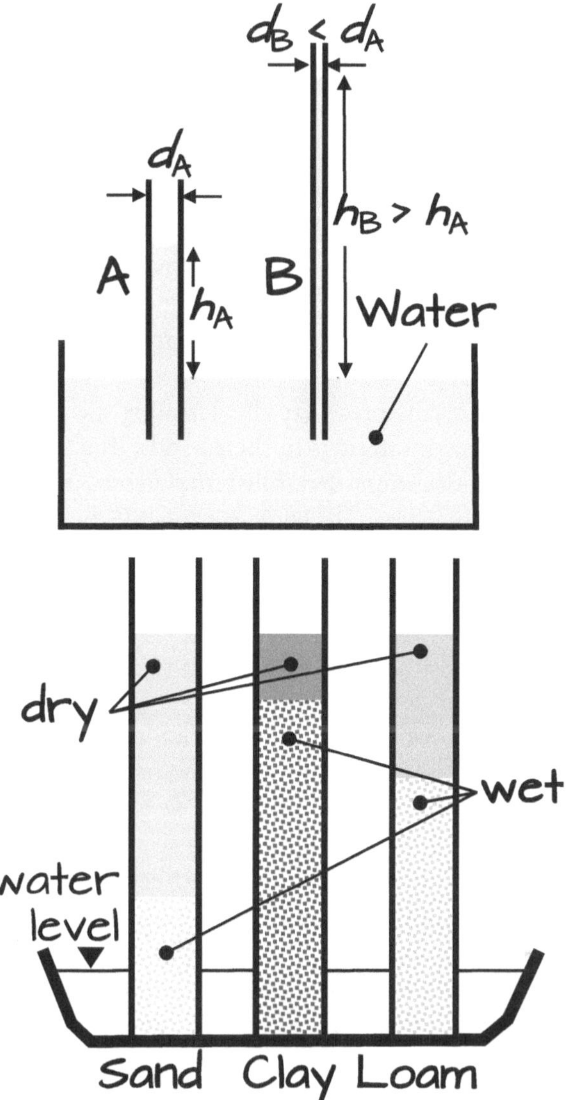

Fig. 2.16 Capillary head

proposed bathing every third day—a gentle reminder of our need to cherish and conserve water. During the worst drought of the twentieth century (1976–1977), British teenagers donned T-shirts with the message "Save water—bathe with a friend" for the same reason. I'm willing to share my choice for the second alternative.

Even if the Vitruvian account of Archimedes' discovery perplexes someone, the miracle of Ganesha, which has surprised millions of Hindus, offers no room for interpretation.

The Ganesha sipping milk miracle occurred on September 21, 1995, when statues of the Hindu deity Ganesha were believed to be drinking milk offerings. A worshipper at a temple in southern New Delhi offered milk to a Ganesha idol. When a spoonful of milk from the bowl was held up to the statue's lips, it vanished, seemingly absorbed by the idol. He woke up the local priest, who performed the test and confirmed the miracle. His scientific approach was sound, as repeatability and reproducibility are critical components of any discovery.

This miraculous event spread rapidly, kindling the hearts of believers who flocked to temples with offerings in their hands. The surge in milk sales mirrored the swell of devotion that filled the streets, transforming the city's rhythm into a flowing stream of faith. Meanwhile, behind this celestial spectacle, scientists delved into the mystery of the vanishing milk. They unraveled the enigma—capillary action, the quiet interplay of liquids in confined spaces, challenging gravity's hold. Demonstrating with dyed milk, they revealed the reality: the statues, with their porous structure, had been absorbing the milk through the subtle magic of capillarity.

Despite ridiculing comments in the media, particularly in Western countries, capillary action remains a marvel of the natural world, and a phenomenon that requires additional investigation. It looks as an intricate dance of adhesion, cohesion, and surface tension—a spectacle where water forms bonds with itself and other surfaces. A glass filled to the brim, carefully holding additional droplets, forms a delicate dome, showcasing water's cohesive power. This dome-like shape forms due to the water molecules' cohesive properties, or their tendency to stick to one another. Cohesion refers to the attraction of molecules to other molecules of the same kind, and water molecules have strong cohesive forces thanks to their ability to form hydrogen bonds with one another.

Water's processes at rest also include adhesion—its interaction with other materials. This explains why droplets linger on a window after rain, pausing in their downward trail. At the center of this display is surface tension, the invisible force maintaining the integrity of water's surface. It is the force that gently curves the surface of water in a cup downwards, forming a soft meniscus at the juncture of water, air, and glass. Droplets that stick to a window after a thunderstorm demonstrate the link between adhesion and cohesion, which explains water viscosity.

Surface tension occurs as a result of cohesive forces. This phenomenon causes a liquid's surface to resist rupture when placed under tension or stress from external forces. Water molecules at the water–air interface will create hydrogen bonds with their neighbors, just as water molecules deeper in the liquid do. Filling a cup or glass with water causes the level to drop somewhat. Observing beyond the surface can lead to a better understanding of the phenomenon. The water's surface tension causes the sinking of the meniscus. The meniscus shape at the water–air interface is caused by the cohesion of water molecules as well as their adhesion to glass.

The capillary action occurs when the adhesion to the container is stronger than the cohesive forces between the water molecules. Inside the narrow space of a straw, water defies gravity, ascending in a delicate equilibrium of forces. The water's concave meniscus, a symbol of adhesion's greater influence over cohesion, subtly defines water's behavior in a narrow tube. As the tube diameter lowers, the relative surface area of the interface within the tube increases, allowing capillary action to push the liquid higher than in larger-diameter tubes.

Yet, this interplay of forces is not fixed. In the slender tube of a thermometer, mercury displays a convex meniscus, a celebration of cohesion's dominance over adhesion. It serves as a reminder that nature is full of exceptions, of wonders yet to be explored. The shape of the water's meniscus is concave because adhesion force of the liquid to the container is greater than that of cohesion. The opposite happens for mercury, when cohesion exceeds adhesion.

Therefore, the story of Ganesha's divine thirst, a wondrous spectacle to the faithful, unfolds its secrets under the scrutinizing lens of science. The statues, with their porous essence and adorned in nature's splendor, drew in the milk through capillary action. It is a physical phenomenon, manifesting in the realm of the divine. In reverence and fascination, we recognize Ganesha's role as the *Remover of Obstacles*, yet through science, we find explanations that anchor the divine in the realm of the tangible.

In the interplay of water, the narratives of ancient bathtubs, and divine miracles, we encounter the poetic language of science. It speaks of wonder and inquisitiveness, bridging the celestial and the earthly. In these age-old stories, whispered across generations, lies the perpetual dance of discovery and awe.

3

Water in Motion

The soul of man
Resembles water:
From heaven it cometh,
To heaven it soareth
And then again
To earth descendeth,
Changing ever.
Johann Wolfgang von Goethe, Spirit song over the waters, 1779

Human Control of Water Flow

Water is a vital component of life. It covers more than 70% of the Earth's surface, is essential to biology, and has always captivated people, who have never stopped questioning why water moves the way it does. The basic observations of water in motion were already documented in the cultural and spiritual traditions of ancient civilizations, even if the word hydrodynamics had not existed for thousands of years. The development of the science of water motion over time required increasingly complex ideas to understand the basic mechanisms. Although it may seem simple, understanding how water flows influence our daily lives is not so easy, from the water we drink to the ships we sail and the weather patterns we experience.

The early civilizations settled near water bodies, understanding their necessity for survival. But there was much more intrigue to the movement, activity, and patterns of water than just consumption. The myths of massive floods, stories about sea gods and goddesses, and the development of early irrigation

R. Rosso, *Five Easy Pieces on Water*, https://doi.org/10.1007/978-3-031-69276-5_3

systems show the amazement and reverence of humans toward water. Without taking into account humanity's emotional bond with the fluidity of existence, one cannot understand water.

As civilizations flourished along the banks of mighty rivers—the Tigris, Euphrates, Nile, and Indus—water became both a deity and a tool. The river provided nourishment and enabled agriculture. One of the earliest accounts of studying water motion comes from Ancient Egypt. With its annual floods, the Nile presented both opportunities and difficulties. Many centuries before Leonardo da Vinci invoked that knowledge about water requires experience first, the Egyptians used the "Nilometers" to gauge the Nile's inundation levels. This gave them the ability to design irrigation systems and canals relying on intuitive understanding of water's behavior. Through their riverine rituals, they came to understand that the Nile's regular flooding was not a random event but rather a celestial dance, driven by the Sun, replicated on Earth.

Similar to this, the Indus Valley Civilization demonstrated a profound understanding of water motion with its amazing drainage systems. Hydraulics was used extensively in the building of stepwells and municipal water supply in ancient Indian cities such as Mohenjo-Daro and Dholavira. The Punjabi Harappan culture has sophisticated knowledge of drainage and water conservation. The Chinese built intricate canal systems as well as the longest canal in the world, the Jing–Hang Grand Canal, demonstrating their mastery of water motion.

The Greeks first established the systematic approach. Archimedes of Syracuse took a legendary bath that led to the discovery of the principle of buoyancy and discussed water in motion in his pioneering treatise. Aristotle wrote a great deal about water motion among other natural phenomena. He understood how rivers shaped the terrain and established the foundation for later studies on the mechanics of moving water. Even if his observations are not always correct by today's standards, they set the path for centuries before his theories were confuted.

The aqueducts built by the Romans are a consistent proof of their engineering brilliance. These imposing structures, built under precise slope constraints, guaranteed a steady supply of water into cities and farms. Vitruvius combined theoretical knowledge with practical engineering awareness in his voluminous treatise on architecture, where several issues of water motion are discussed, including aqueducts and water wheels. Hydrodynamic expertise was entrusted to the monks of medieval Europe, which were frequently located next to placid lakes or gently flowing rivers. Monastic manuscripts occasionally delved into the enigmatic aspects of water, encompassing both spiritual and scientific aspects. In the ninth century, the Arab scholar Al-Jahiz wrote detailed observations on water properties and its interaction with various materials.

Long before formal theories were developed, six centuries later, Leonardo da Vinci's careful drawings show how controlling canals to convey and share water among different stakeholders, how push wheels for milling grain or pumping water, how regulating navigation. He also revealed the vortex nature of water, providing a glimpse into turbulence, observed and described the streamline effect, focused on how water flows around obstacle to understand the influence of solid structures.

Conservation of Mass

The basic conceptualization of water flux was first proposed by Heron of Alexandria two millennia ago. He established a relationship between the cross-sectional area of a spring's outflow and its discharge, as we call the volume of water passing through a section per unit time. For a pipe or a river, the discharge is the rate of volumetric flow or how much water is moving past a point in a given time interval. It is commonly measured in cubic meters per second using SI metric.[1]

Heron moved by one of Archimedes' findings, stating that "water flows faster and faster when it is forced through a bottleneck, and the increase in water velocity is proportional to the narrowing of the passage channel section." Accordingly, the discharge is the product between cross-sectional area and flow velocity. If one must convey a water flux with constant discharge, this product must be constant. This concept was embedded in the capability of ancient civilization to build canals using stone, wood, concrete, and masonry; as well as the lead pipes that one finds in Pompei's spas and villas. It also paved the way for the development of the core idea of continuity.

The Heron's insight is the paradigm for the first basic concept of hydrodynamics, which was introduced in 1639 by Benedetto Castelli, a close friend of Galileo Galilei, his guru. One of the long-standing mysteries of scientific inquiry is why Heron's discoveries were not widely accepted. Although no ancient treaties mention it, the Romans who constructed aqueducts, irrigation canals, and spas two millennia ago were undoubtedly aware of the

[1] The International System of Units, internationally known by the abbreviation SI (from *French Système International d'Unités*), is the modern form of the metric system and the world's most widely used system of measurement. SI units include: s, second, for time; m, meter, for length; kg, kilogram, for mass; K, kelvin, for thermodynamic temperature; mol, mole, for the amount of substance; and A, ampere, for electric current.

relationship between cross-sectional area and flow velocity. We can only speculate that Heron's outright rejection of the idea of a vacuum—an essential issue to Aristotle's philosophy—was detrimental to him. It is possible that Aristotelians' ostracism prevented them from recognizing Heron's work, given that Aristotle had a significant impact on scientific thinking up until the Renaissance in Western countries.

Does moving water maintain its mass? The response is unquestionably affirmative since Heron was not the first ancient Greek philosopher to recognize that nothing originates from nothing. What is mass conservation's result? The principle of continuity aids in the comprehension of water flow. Assume you using a garden hose. Water pours out when it is turned on. Unless there is a leak, the amount of water entering the hose and the amount of water exiting must be the same. This is the basic idea of continuity: providing nothing is added to or removed from the flux, the amount of water entering and exiting must be identical.

If a hose has a thin section, water moves more quickly through that section. Why? because a smaller area must be traversed by the same amount of water. The water slows down as the hose widens. This picture relies on three assumptions: that water is incompressible, that its density is constant, and that its flow is steady; it does not vary over time.

In general, water is essentially uncompressible, especially under typical conditions. Squeeze a balloon that is filled with water, if you will. The balloon's shape changes, but the volume of water inside does not; it does not condense into a smaller space (Fig. 3.1a).

Constant density for water means that, in normal circumstances, the volume of water in a particular space stays pretty much the same. Density is a measure of the quantity of material packed into a unit volume. In the case of water, this means that the volume of water in a given volume (such as a cup) stays relatively constant regardless of how much you pour into a container. Regardless of location or type of container, the mass of one liter of water is always roughly the same (Fig. 3.1b). In practice, water density does not vary significantly with mild temperature fluctuations until it is very near the threshold at which it turns into ice. However, water in nature contains solutes that modify its density. How much depends on type and concentration of solutes.

Simply put, the continuity principle says that whatever enters a system must exit it unless it builds up inside. As introduced above, water's density is almost constant for an incompressible fluid, such as water flowing in a river (Fig. 3.2) or running through a conduit (Fig. 3.3). Therefore, if there is no lateral inflow or outflow, the discharge at any two sections along the flow

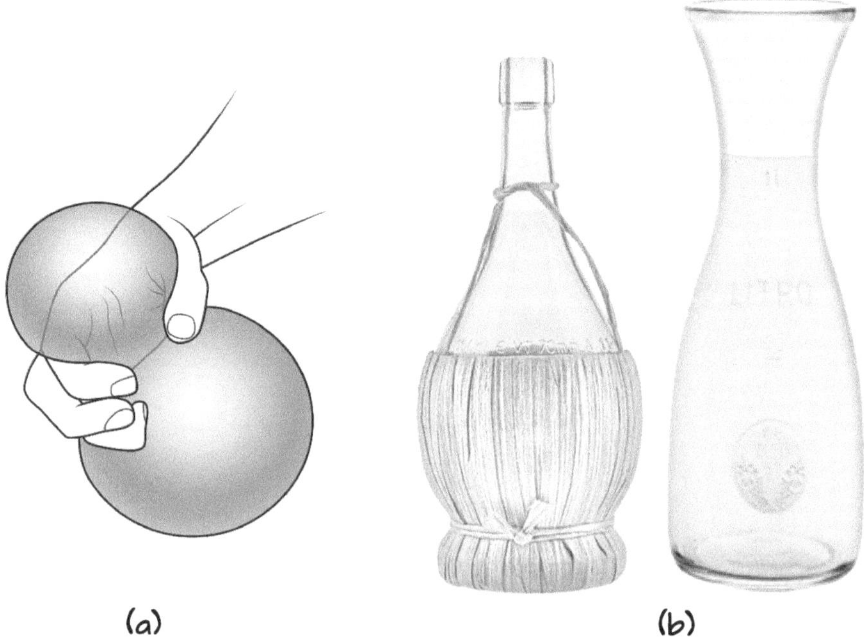

Fig. 3.1 Incompressible fluid, unsqueezable (a) and shapeless (b)

direction is the same, and the product of the wetted area and flow velocity will not change: $A\ V = constant$. The wetted area is measured by the fluid cross-section perpendicular to flow direction. In order to maintain a constant volumetric flow rate, velocity V increases with decreasing wetted area A, and vice versa. Steady flow is a further implicit assumption. To account for time variations, more complex formulations of the continuity equation would be needed in real-life scenarios with unsteady flows.

Conservation of Momentum

Momentum is a basic concept in hydrodynamics that describes how water, or any other fluid, moves. It is basically the product of the fluid's mass and its velocity. In hydrodynamics, momentum is not just a measure of how much fluid is moving but also how fast it is moving, encapsulating both the quantity of the moving water and its speed in a single concept.

The study of water motion requires to introduce the principle of momentum conservation. It states that the total momentum in a system stays constant until acted upon by outside forces. This concept aids in forecasting the

Fig. 3.2 Continuity in open channel flow

streamflow behavior in the presence of obstructions, changes in channel geometry, or variations in elevation and slope in rivers, channels, or any other hydraulic system.

For example, as water flows through a river bend, its momentum keeps the water flowing in a straight path, which causes the outer bank of the bend to erode more deeply and at higher velocities. Using their knowledge of momentum, engineers create structures and riverbanks that either control or resist the flow of water's inherent tendency. Alternatively, we can accommodate nature-based solution because

> the parallelism of the directions, and the rectilinear shape of the riverbed cannot naturally take place except under the assumption of the homogeneity of all parts of the bottom. Considering the riverbeds, as they really are, intertwined with boulders, gravel, sand, clay, and other materials that are otherwise tenacious, one will have the reason, for which the streamflow is now approaching one, now

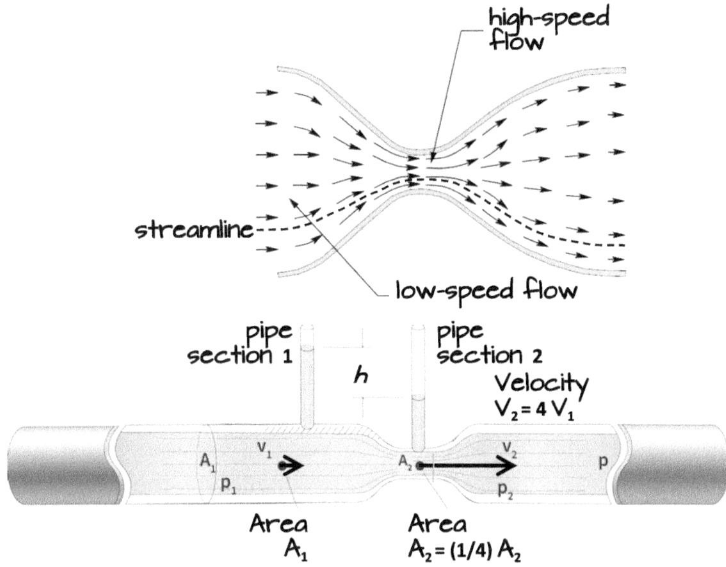

Fig. 3.3 Continuity in pressure flow

> to the other bank, and the rivers in their length present us with a series of concave, and convex arches.[2]

The conservation of momentum is frequently employed in engineering applications in conjunction with the equations for mass balance and energy conservation to resolve challenging flow problems. These include estimating the stress put on hydraulic structures, forecasting how water will flow in pipelines and open channels, and creating effective water management and transportation systems. Furthermore, assessing the effects of abrupt changes in water flow on infrastructure like dams, spillways, and bridges as well as studying dynamic events like hydraulic jumps—which involve rapid changes in depth and velocity—needs an understanding of the momentum conservation principle.

Momentum measures the quantity of motion of the object. It is a measure of the inertia of a moving object. In hydrodynamics, momentum is currently evaluated using the unit of kilogram meter per second (kg·m/s) in SI units. This unit of measurement gives the amount of motion that a fluid possesses

[2] Frisi, P. (1777). *Istituzioni di meccanica, d'idrostatica, d'idrometria e dell'architettura statica, e idraulica ad uso della regia scuola eretta in Milano per gli architetti e gli ingegneri*, Milano: Galeazzi (*Principles of mechanics, hydrostatics, hydrometry and static architecture, and hydraulics…*, in Italian).

by directly reflecting the product of mass (measured in kilograms) and velocity (measured in meters per second).

Do not mix up the concepts of moment with the force. According to Newton's second law of motion, force is the product of acceleration and mass when expressed in SI units, $1\ N = 1\ kg{\cdot}m/s^2$. It explains the push or pull applied to an object that accelerates it. Since force and momentum describe separate concepts, they are measured in different units even though they are connected.

When water flows away from a point, there must be a decrease in the amount left behind due to continuity. If flow velocity is described more accurately in 3D as a vector V, the mass flowing in a unit time across a unit wetted area is the component of the product ρV normal to flow direction, with ρ denoting water density. As a result, the velocity variation must match the density's temporal variation. When density is constant, there is no change in flow velocity V over time: **div** $V = 0$ in mathematical form, where "**div**" is the divergence operator, namely the sum of the rates of variations of flow velocity along the three Cartesian axes. Leonhard Euler presented his equation for an incompressible fluid with a constant and uniform density at the 1752 session of Berlin Academy.[3]

Water has mass because it is composed of individual particles. These particles accelerate in response to external forces. This concept is crucial for understanding water flow and pressure changes in hydrodynamic systems. Returning to the elementary example of water running through a pipe: water accelerates because of the change in wetted areas as it reaches a narrowing in the pipe. This can be understood by connecting Newton's second law with the concept of mass conservation.

Driving Forces

The primary force controlling water motion on Earth is gravity.

> On how one cannot describe the behavior of water, if one does not begin by defining what gravity is and where such gravity is generated and where it is cancelled.[4]

[3] Euler, L. (1757). Principes généraux du mouvement des fluids. *Mémoires de l'académie des sciences de Berlin*, Vol.11: 274–315 (The General Principles of the Movement of Fluids, in French).

[4] Leonardo Da Vinci, *Leicester Codex*, Folio 26v, 1504–1508.

As a result of the pressure gradient it produces, water moves from higher to lower elevations. Conversely, buoyancy refers to the upward force that water applies to an object submerged in it, enabling it to either float or sink. The cohesive forces that exist between water molecules at the air–water interface give rise to surface tension. It creates a number of phenomena, including droplet formation and capillary action. Surface tension also influences wave motion and pattern formation on the water's surface.

The Coriolis effect, triggered by the Earth's rotation, causes moving objects, such as water currents, to be deflected to the left in the Southern Hemisphere and to the right in the Northern Hemisphere. Large-scale atmospheric and oceanic circulations are influenced by this effect, which gives rise to the trade winds and the Gulf Stream. Depending on which hemisphere you are in, you may see the Coriolis effect in action when water flows down a bathtub drain in opposite directions. If that were the only force acting on a current, in Germany it would rotate clockwise while in Chile it would counterclockwise, while at the Equator it would not rotate at all.

In practice, other effects play a role to drive vortex rotation. You need a perfect tub since with regular tubs, things like a slight asymmetry in the drain's form will decide which way the circulation goes. Even in a tub with a rightly symmetric drain, the circulation direction will be primarily influenced by any residual currents in the bathtub left over from the time when it was filled.

The groundbreaking work of Sir Isaac Newton revolutionized our understanding of motion, including the principles governing fluid dynamics. Newton's second law of motion, which connects force, mass, and acceleration, provided a framework for understanding the complex interactions between water and outside forces.

The law of inertia, which is Newton's first law, states that unless an outside force acts upon an object in motion, the object will tend to stay in motion. This principle helps elucidate the tendency of water to flow in a continuous way, maintaining its motion until influenced by external factors such as gravity, wind, differences in pressure and density. In the ocean, for example, water masses will maintain their motion unless they are affected by outside factors such as wind, topographical changes on the sea floor, or gradients in salt and temperature.

To measure the impact of external forces on fluid motion, Newton's second law establishes a relationship between force, mass, and acceleration. This law states that an object's acceleration is inversely proportional to its mass and directly related to the net force exerted upon it. Accordingly, the force acting on an object is equal to its mass m times its acceleration $\boldsymbol{a}$, or $\boldsymbol{F} = m\boldsymbol{a}$. In water motion, this principle helps explain how changes in forces—like wind stress on the ocean surface or gravitational pull on river water—can accelerate running water and change its flow patterns. For example, wind creates a force that can initiate and modify currents when it blows across the ocean's surface. Gravity causes rivers to flow downhill.

Newton's Second Law is reformulated in hydrodynamics to take into consideration the fluid's flow and the continuous distribution of mass. The result is the principle of conservation of momentum and the Euler's equation. The force acting on a water element results not only from external sources, like gravity, but also from pressure gradients, viscous stresses, and other fluid properties. The Navier-Stokes Equations provide the most thorough expression of Newton's Second Law in hydrodynamics. Their complex mathematical structure accounts for the combined effects of pressure, fluid velocity, viscosity, and external forces.

Navier-Stokes Equations explain how changes in pressure and fluid velocity lead to the acceleration of water. For example, an imbalanced force caused by the pressure differential and velocity gradient causes the water in a river to accelerate as it narrows. Navier-Stokes Equations explain how the viscosity of water creates internal friction that impacts the flow, particularly when the water is moving slowly or if flows as a shallow water in small channels. These equations incorporate forces like as gravity and wind pressure on the ocean surface, enabling to study how these factors affect water motion in rivers, lakes, and other water bodies.

The reciprocal relationship between the forces that fluids exert was underlined by Newton's third law, sometimes known as the law of action and reaction. This principle is crucial in understanding the complex interactions between water and submerged objects, such as ships or aquatic organisms. For every action, there is an equal and opposite reaction. In hydrodynamics, this is observed in wave motion. As wind blows across the ocean's surface, it pushes against the water (action), and the water pushes back against the wind (reaction) forming waves. Comparably, when water hits a barrier or bounces off a riverbank, the reaction of the water reflects this interaction, often causing eddies or changing flow direction.

Conservation of Energy

The law of conservation of energy explains the nature of energy. It asserts that energy can be neither created nor destroyed, but only converted from one kind of energy into another. In other words, the overall energy of an isolated system stays unchanged. Fluid mechanics operates on the same premise with the Bernoulli principle, which explains how energy is conserved in perfect (ideal) fluids. Since the pressure that a fluid exerts is inversely proportional to its velocity in a horizontal flow, Bernoulli proved that the total mechanical energy of a liquid flux remains unchanged at any point along a streamline, assuming that the fluid in question is incompressible and has zero viscosity.

Pressure flow plays a critical role in understanding the behavior of water in motion. Pascal's law, named for the French mathematician and physicist Blaise Pascal, is the fundamental principle of pressure flow. According to this law, pressure is transferred equally in all directions when it is applied to a confined fluid. For instance, this indicates that water applies pressure to the base and the walls of basins equally on all sides.

The energy associated with water pressure is referred to as the pressure head. Usually, it is taken as the height of a water column that would apply the same amount of pressure. Pressure head is the ratio between pressure p and specific weight, given by the product of its density ρ and the acceleration g due to gravity: $H_p = p/(\rho g)$. For instance, pressure head H_p provides a convenient way to quantify the potential energy associated with pressure flow in a reservoir. It makes it easy to compare pressure levels at various reservoir locations.

Conservation of energy explains how pressure and velocity affect water motion. The essential principle relating to water's pressure, velocity, and elevation in a steady flow is known as Bernoulli's equation:

$$p + \frac{1}{2}\rho V^2 + \rho g z = \text{constant},$$

where p denotes the pressure, ρ is the fluid density, V is the velocity, g is the acceleration due to gravity, and z is the elevation. This constant is called the hydraulic load.

By introducing the concept of streamline, this theory becomes clearer. These curves show the direction of motion of a fluid element at any point in time (Fig. 3.3). If you throw a leaf in a stream, a streamline created is the path of the leaf as it floats downstream. Of course, the leaf can follow countless routes depending on where it lands in the stream after contact. Streamlines exist underwater as well. Imagine a neutrally buoyant particle that is immersed and neither sinks nor floats. This particle travels down the river in a streamline, like a wet twig.

When values for velocity, pressure, and elevation are plugged into Bernoulli's equation, the result of the equation will be the same (constant) at every point along the streamline. Accordingly, the fluid's velocity increases when its potential energy or static pressure decreases at the same time. Swiss mathematician and physicist Daniel I. Bernoulli published these findings in his book

Hydrodynamica, sive de viribus et motibus fluidorum commentarii, published in 1738. This knowledge is actually the outcome of a French-Swiss team effort including Leonhard Euler, Bernoulli's co-worker and friend, along with Francois Alexis Claude Clairaut and Jean le Rond d'Alembert.

These findings offered a more thorough and organized comprehension of prior intuitive knowledge. For example, the concepts of hydraulic head and the outflow law from a tank orifice are presented in the volume *De motu aquarum* of his treatise *Opera geometrica*, published in 1644 by Evangelista Torricelli, a student of Benedetto Castelli, in turn a pupil and collaborator of Galileo Galilei (Fig. 3.4). In addition to confirming Heron's intuition on the link between wetted area and flow velocity, Torricelli proved that the rate of efflux from a hole was proportional to the square root of the height H of the water column load above the opening. The Bernoulli principle has further proved that $V = \sqrt{(gH)}$. Although only the Bernoulli equation has been able to prove it, this equation is known as Torricelli's law.

The Bernoulli's equation shows how pressure, velocity, and elevation in a water system interact. Imagine a pipeline that connects a water reservoir at the summit of a mountain to another reservoir at a lower elevation downstream, where a hydroelectric plant or an aqueduct tower is located. Water in the top reservoir travels slowly toward the bottom opening at zero or very small velocity. The pressure is likewise very low for a water particle close to the top of the reservoir. On the other hand, the reservoir elevation is much higher than that

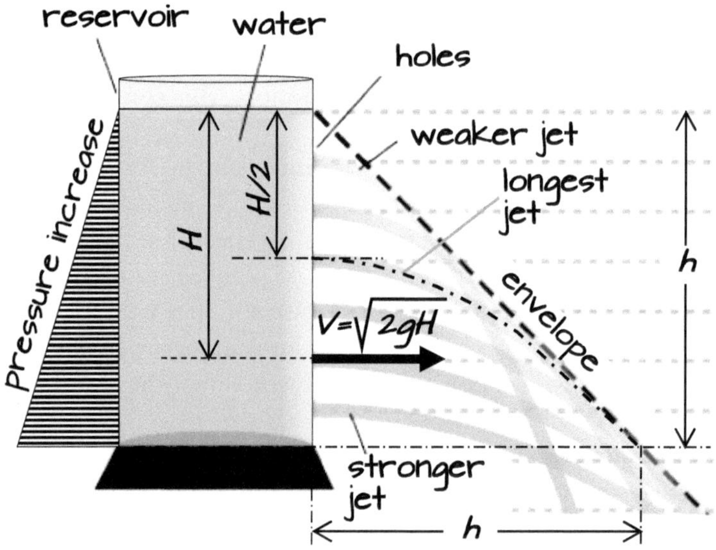

Fig. 3.4 Torricelli's law

of the bottom one. Let us now track a drop of water as it outflows from the reservoir into the pipeline. As it moves downstream, its elevation decreases, velocity increases, and pressure rises since water moves via an enclosed piping system. The Bernoulli's approach allows to assess the amount of energy available for supplying the municipal water distribution system or the hydroelectric power plant downstream.

The energy associated with water pressure is referred to as the pressure head. Usually, it is defined as the height of a water column that would apply the same amount of pressure. As introduced above, pressure head H_p is the ratio between pressure p and the product of its density ρ and the acceleration g due to gravity: $H_p = p/(\rho g)$. Since pressure head is measured in units of length, such as meters, it offers a handy approach to estimate the potential energy associated with pressure flow at various reservoir depths. The Bernoulli theorem allows for assessing pressure levels at different points within a water system when written as

$$H = V^2 / (2g) + p / (\rho g) + z = \text{constant, or}$$

$$H = H_V + H_p + z = \text{constant}$$

where total energy is represented as total head H that equals the sum of three components: kinetic energy expressed by kinetic head $H_V = V^2/(2\,g)$, pressure energy expressed by pressure head $H_p = p/(\rho g)$, and potential energy expressed by elevation z. When water flows due to gravity, its energy is split among these three different components under the assumptions of steady motion, constant density, and perfect (ideal) and incompressible fluid (Fig. 3.5).

As soon as the outlet gate opens and water begins to flow, the dissipative processes cause some of the water's initial potential energy to be lost. The flow

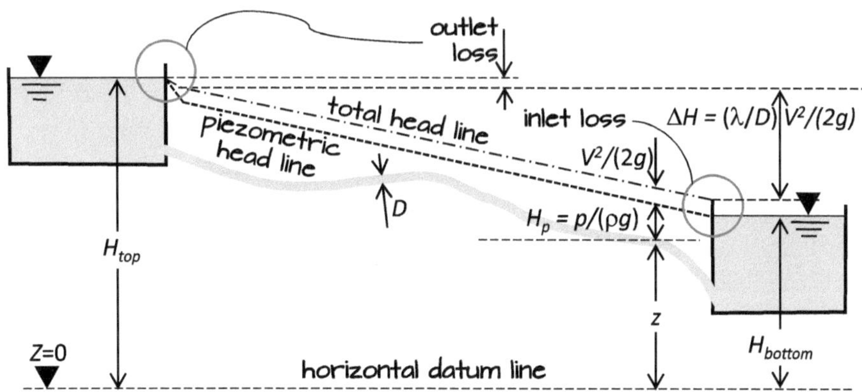

Fig. 3.5 Bernoulli's principle in real world

is then impacted by water viscosity, friction against the duct walls, and irregularities along the path. A portion of the total energy *H* of the still water in the top reservoir is not converted into kinetic or pressure energy as expected under the ideal fluid assumption, it is used to overcome all these resistance factors. As a result, the velocity and flow kinetic energy differ from what the perfect fluid conjecture predicts.

Paolo Frisi, who was named the Italian Newton by D'Alembert, criticized the Franco-Swiss school's purely mathematical approach due to its inability to provide direct practical results based on its theoretical model. Frisi moved in the footsteps of Leonardo and Galileo to claim the impossibility of discovering phenomena in Hydraulics through mathematical analysis, before interpreting them with the understanding of physics, which is a part of the field of hydraulics. Indeed, some compromise is needed.

The actual kinetic head at the bottom reservoir plant, as seen from the reservoir, would be significantly lower than what Bernoulli's calculation indicated. This is accounted for by the concept of head loss, which is the energy in a moving fluid lost as a result of turbulence and friction in the water throughout its journey between the two reservoirs. Pipe length, diameter, and roughness all affect head loss, as well as bends, fittings, and valves.

Turbulence is a major cause of head loss. When turbulence is triggered by some conditions of flow motion, the particles of water move chaotically or randomly. In the turbulent zone, the streamlines continually and quickly change shape and direction so that a particle tracks becomes unpredictable. Turbulent zones take energy from the fluid so contributing to energy losses. One could thus try to minimize turbulence in piping systems, but pipe flow in conduits is usually turbulent in actual water supply and energy systems. Turbulence is also produced by all piping system fittings, including unions, tees, and valves.

Generally speaking, head loss is taken to be proportional to the square flow velocity. In practice, all local and distributed head losses are expressed in terms of units of the kinetic head. When a gate valve is fully open, head loss is around 15% of its kinetic head, while at the inflow gate of the bottom reservoir equals the kinetic head. The square velocity, as well as length and roughness of the conduit dictates the distributed frictional loss along the path.

Can we have a better understanding of turbulence's mechanisms and occurrence? Werner Heisenberg, who was awarded the Nobel Prize in 1932 for his contributions to quantum mechanics and the uncertainty principle, famously declared that if he could ask God two things, he would ask him, "Why quantum mechanics? And why turbulence?" He felt fairly certain that God could

Fig. 3.6 Vortex patterns

respond to the first query only. Although the quote may be apocryphal, Heisenberg indeed banged his head against turbulence for several years.

Laminar and Turbulent Flow Regimes

Have you ever observed a river's streamflow? Water motion looks like a dance that water performs from a peaceful waltz to a wild tarantella. Back in the nineteenth century, Sir Osborne Reynolds explored this dance. He conducted a unique experiment that clarified the flow characteristics of water currents in rivers and pipes. Leonardo da Vinci's sketch of water exiting a square orifice into a pool (Fig. 3.6) was "perhaps the first use of visualization as a scientific tool to study turbulent flow".[5] Reynolds' approach transformed our understanding on the world by transposing Leonardo's vision into a scientific experiment.

When water moves, it can do it in two ways: calm or crazy. Imagine a group of people walking in a straight line without bumping into each other, like the Japanese entering to the theater. This is a bit like calm flow, which

[5] Gad-El-Hak, M. (1998). Fluid Mechanics from the Beginning to the Third Millennium. *International Applied Mechanics*, *14*(3), 177–185.

hydrologists call "laminar." But then, picture a crowd of people rushing around, colliding, and fluctuating in all directions. This is more like crazy flow, or "turbulent," like Italians queuing at the box office of the Theater at the Scala, the historic opera house in Milan, Italy. Reynolds wondered, "What makes liquids switch from calm to crazy flow?" To solve the puzzle, he did experiments with his glass tube and colored water.

Imagine a glass pipe that lets you see what's happening inside. Reynolds connected the pipe to the outlet of a water tank kept at a constant level, observing the continuous outflow. Additionally, he installed a nozzle at the pipe's inlet to inject colored water into the base flow. This way, he could watch how the water flowed under different boundary conditions (Fig. 3.7). He varied flow velocity by changing the level in the tank, all while keeping an eye on the glass tube. This allowed him to piece together the puzzle of what factors triggered the transformation from one flow style to the other. Reynold's technique explored in the laboratory what Leonardo da Vinci had somewhat pioneered.

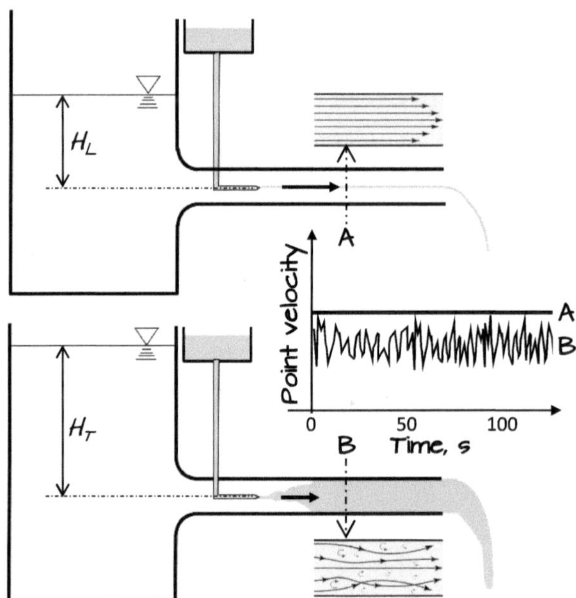

Fig. 3.7 Reynold's experiment

> It is also astonishing that Leonardo should have used for his studies of the motions of water, colored water in which the different streams could be distinguished.[6]

Water molecules move smoothly following a predictable path when the flow is laminar. Conversely, water turns into a maelstrom of twists and turns in turbulent flow, as if each molecule decided to move independently. Reynolds introduced a number that could predict when water would go from calm to crazy. It is known as the "Reynolds number," or simply *Re*. Think of the Reynolds number as a note on the musical scale that water can hear. It depends on three key factors: how fast the water flows, how thick the water is, and the cross-section size of the flow, which is the diameter of the pipe in pressure flow. The Reynolds number is a dimensionless quantity that depends on these three factors, $Re = VD/\nu$, namely flow velocity V, pipe diameter D, kinematic viscosity ν. When a certain note is played, a calm water flow turn to be crazy.

There is a special value of this number, called the "critical Reynolds number," a gateway that switches the transition from laminar to turbulent flow. When the Reynolds number exceeds this value, the flow changes from calm to crazy. It is like someone flicking a switch, and suddenly the water goes from behaving politely to getting all jumpy and wild.

If water could listen to a pianist who plays the entire A major scale on the keyboard, after he presses the A4 musical note key on the central octave of the piano, which corresponds to the 440 Hz frequency, water's dance would abruptly switch from a slow waltz to a boogie-woogie. In this case, Re = A4. As a rule of thumb, laminar flow typically occurs when water is flowing at a Reynolds number lower than two thousand. The flow becomes turbulent as the Reynolds number rises above four thousand. Transitional water flow occurs in the middle (Fig. 3.8).

Laminar flow in rivers is typically observed in shallow, leisurely flowing streams with no whirlpools or eddies. This is more common in the upper reaches near a glacier spring or in vast river courses where the water flows slowly and effortlessly through an unobstructed path. As the river's slope increases or when it encounters obstacles like boulders and rocks, the flow becomes turbulent, displaying chaotic, swirling motions as seen in rapids or waterfalls. River flow is typically turbulent, while it occasionally becomes laminar in the inflow reach to a lake or in swamps (Fig. 3.9).

[6] See: Gille, B. (1966) *Engineers of the Renaissance*. Cambridge: The MIT Press, p. 181.

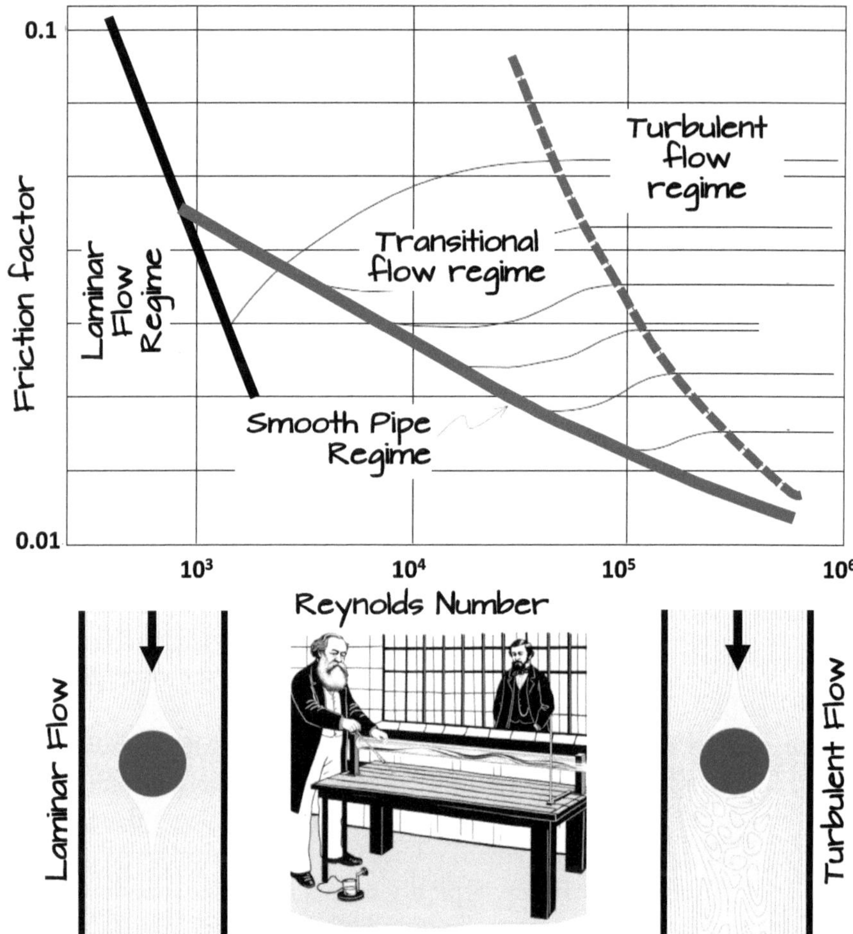

Fig. 3.8 Laminar and turbulent flows

In lakes, laminar flow is observed in undisturbed areas, especially near the edges where the water is calm and there are no significant winds or currents. Deep currents may be laminar. Wind frequently creates turbulent flow in lakes and reservoirs, which results in waves and surface currents. This is most noticeable during storms or heavy winds, as the water's surface becomes erratic and chaotic.

Laminar flow is uncommon in the wide open ocean because of the continuous movement triggered by winds, tides, and currents. But occasionally, it can be seen in quiet, extremely deep waters. The ocean is predominantly characterized by turbulent flow, visible in the form of waves, tides, and

Fig. 3.9 Laminar and turbulent channels

whirlpools. Both surface and deep ocean currents display turbulent characteristics.

The decision between turbulent and laminar flow in man-made systems is based on the particular needs of the process. Laminar flow is frequently the best option for reducing friction and preserving exact control because of its smooth, organized motion. However, turbulent flow is favored for operations that have a need for the efficient mixing, aeration, or quick heat transfer, even if the system is more chaotic and energy-intensive.

Laminar flow is frequently used in wastewater treatment facilities for procedures requiring a slow, continuous flow of water, such as in certain types of sedimentation tanks or in filtration systems where the water must flow through a medium without causing disruption or channeling. Conversely, at different phases of water treatment, such as in aeration tanks where the objective is to enhance water mixing and oxygenation, turbulent flow is crucial. It is also common in rapid mixing chambers where chemicals are added to the water for treatment purposes.

Blood is a liquid that flows through our veins and arteries, but it is much more viscous than water. However, we know more about how blood flows through our bodies thanks to Reynolds' ideas. Through an understanding of the behavior of blood at different flow rates, if it flows quickly or slowly, medical doctors can better treat heart and circulation problems, ensuring our body stay healthy and balanced. Reynolds provided an understanding of Leonardo da Vinci's groundbreaking discoveries about blood circulation, let alone turbulent fluxes, vortexes, and eddies.

Eulerian and Lagrangian Views

Understanding how water moves is important for a variety of tasks, such as managing water resources and forecasting weather and ocean behavior. Every water flow is unique, and a vast amount of specific knowledge has been produced to investigate it. As Leonardo da Vinci argued, observation is essential to understand water. Water motion can be observed from two distinct viewpoints: static and cinematic. As a result, the Eulerian and Lagrangian approaches—two separate but related strategies—emerged. The first one offers special insights into the behavior of water since it is consistent with the model that Euler employed to investigate water motion. The second one bears the name of the Italian-French mathematician Joseph-Louis Lagrange, who approximately fifty years later provided further fundamental contributions to our understanding of water (Fig. 3.10).

Herman Hesse's *Siddhartha,*[7] a novel of spiritual discovery, helps introducing these two opposing viewpoints. It tracks the protagonist Siddhartha on his quest for enlightenment. The river, an enduring presence and a symbol of life's ever-changing yet eternal essence, is central to the story. This river, with its depths of wisdom and its flowing waters, serves as an appropriate metaphor for the two viewpoints—the Lagrangian and the Eulerian approaches.

OpenAI. (2024). ChatGPT (4) [Large language model] https://chat.openai.com

Fig. 3.10 Eulerian and Lagrangian views

[7] Hesse, H. (1922). *Op. Cit.*

In his pursuit of knowledge, Siddhartha takes on a variety of vocations, including that of an ascetic, a merchant, a lover, and eventually a ferryman. From a Lagrangian perspective, he moves across various roles like a particle of water throughout a river. From this angle, one can track a single particle as it travels through a fluid, overcoming obstacles and undergoing changes. The distinct and illuminating experiences of Siddhartha reflect the individual path of a water particle. Similar to how Siddhartha's interactions and experiences shape his journey toward enlightenment, a particle's trajectory is determined by its interactions with its environment, its dance with other particles, and its response to natural forces.

In Hesse's story, the river is more than just a setting; it is a character in itself, symbolizing the interdependence of all things and the eternal flow of existence. After becoming a ferryman, Siddhartha spends his days attempting to understand the river's secrets by listening to it. He is gazing fixedly at the river, trying to make sense of its moods, rhythms, and patterns at particular points in time. This is similar to Eulerian method. Here, one watches the fluid's behavior at fixed spots rather than tracking the path of a particle. It is about figuring out the patterns, the ebb and flow, and the group dynamics at specific locations. Similar to the Eulerian viewpoint, Siddhartha gains his insights from the river as he listens to its murmurs and studies its nature through this immovable observation.

By the end of the book, Siddhartha understands that enlightenment is achieved by a holistic view rather than by a single method. The lessons of the river are deep because Siddhartha gains knowledge from both his journey (Lagrangian) and his sedentary contemplation (Eulerian). Similar to this, a thorough comprehension of hydrodynamics results from an appreciation of both the individual trajectories and the collective behavior. In Hesse's *Siddhartha*, the river skillfully demonstrates this dichotomy, highlighting the idea that true insight comes from fusing unique experiences with life's larger rhythms.

Imagine you're watching a river run beneath you while standing on a bridge. The Eulerian approach involves concentrating on a particular spot or region inside the river and observing the behavior of the water as it passes by. It is similar to putting up a camera to capture a single image of the river. You observe many water droplets move across your range of vision, yet your gaze is fixed on the same spot. This method works quite well for figuring out the pressure, velocity, and flow characteristics at specific river locations. This includes on-site measurement as well as remote sensing for estimating of river discharge.[8]

When attempting to understand the fluctuations in water velocity and discharge, as well as pressure and temperature at various places along the river's path, the Eulerian technique serves as our guide to river dynamics. Imagine a scenario in which our goal is to assess how the construction of a dam affects the downstream streamflow. Through this approach, we can measure the changes in velocity and turbulence at various locations, enabling the ability to predict potential impacts on riverbed erosion and aquatic habitats.

Now picture yourself floating down a river in a boat rather than standing on the bridge. The Lagrangian method involves tracking a single water particle—in this case, your boat—as it travels down the river. It is similar to mounting a GPS tracker on the boat and tracking its route. You watch as that specific particle (or boat) moves, noting changes in its speed, direction, and surroundings. This method helps you comprehend the water motion from the viewpoint of the individual particle by providing you with a sense of the journey taken by each one.

While not as precise as a measurement tool like a current meter or a weir, the float method (Fig. 3.2) can offer a reasonable approximation of flow velocity and its associated discharge if the wetted area is measured as well. Measuring how long a floating object takes to travel a specified distance downstream is the basic idea, although monitoring surface velocity can be also performed by other, more sophisticated methods, including remote sensing from satellite imagery. Using a correction factor, the stream's mean velocity is found. In a nutshell, this technique consists in a floating item to measure the water's surface velocity, which is then multiplied by the channel's width and average depth. More accurate measurements can be performed by injecting a tracer to be sampled at different downstream river cross-sections after the upstream initial release.

This method enables comprehension of the movement of sediment and the dispersion of pollutants in river systems. I used nuclear tracers to investigate the trajectories of gravel and boulders along a mountain creek fifty years ago. Think about an example where there is an upstream chemical leak. We may closely monitor the contaminants' route by inserting innocuous tracers that mimic their behavior. With this information, we may more accurately forecast the potential fate of the contaminants and develop plans to mitigate the impact of pollutants on ecosystems and people living downstream.

As we explore the frontiers of scientific knowledge, it becomes clear that the Eulerian and Lagrangian methods are not mutually exclusive but rather

[8] Masafu, C., Williams, R., & Hurst, M. D. (2023). Satellite video remote sensing for estimation of river discharge. *Geophysical Research Letters*, *50*, e2023GL105839.

complementary facets of the unified quest for knowledge. Think about how these approaches could be used to explore the intricate networks of urban rivers and canals. Through the integration of Eulerian insight into flow velocities and Lagrangian particle tracking, the transport of pollutants can be holistically examined, flood risks can be assessed, and riverbank protection plans can be devised. Furthermore, technology has given us strong instruments to combine these strategies. Sophisticated numerical models enable the simulation of both perspectives, creating a realistic picture of water movement. The combined approach is also useful to analyze water supply systems, as it occurs in aqueduct assessment, where we can track particle mobility and mimic water flow to optimize water distribution and reduce losses.

Open Channel Flow

One of the enduring problems in water motion is understanding open channel flow. Unlike water in a pipe under pressure, it happens when water flows in a natural or constructed channel with a free surface exposed to air. This includes the spillways that shield our towns from flooding, the canals that feed agricultural areas, and the rivers that carve through landscapes.

At the heart of open channel flow are the two fundamental concepts of energy and momentum. Let us start by discussing energy. The main concept here is specific energy, which is essentially a measurement of the work that the water can accomplish as a result of its position and speed. Imagine a stream of water cascading down a mountainside. Gravity pushes it along as it hurtles downhill, providing it with the energy to shift boulders, carve gullies, and sculpt the terrain. This energy depends not just on the water's speed but also on its depth. The energy profile of a deep, slow-moving river differs from that of a shallow, swift-moving mountain creek.

But how does water choose its path? That is where momentum comes in. Momentum is that keeps water moving in the same direction. When a stream drains a mountainside, it wants to keep going straight because gravity dictates following the topographic gradient, the maximum slope track. But, the water's momentum can alter the path, influencing its flow and energy, if the land's slope changes, the river bends, or the channel narrows.

This process is mostly driven by gravity, which causes the water to flow from higher elevations to lower ones. However, the channel's roughness, form, and local slope all have an impact. A steep slope can turn a tranquil stream into a torrent, while a smooth channel allows water to flow more easily than a rough, rocky bed.

> At any place, then, as water becomes deeper it tends to flow faster. When it moves downhill, water acts like any other body that is moved by gravity. It would move ever faster, like a ball rolling downhill were, it not held in check by friction against the channel bed and banks. The speed at which water moves is a balance between gravity and friction. But as the water in any natural stream gets deeper, the area against which the water rubs does not substantially increase. For this reason, gravity becomes more important to water velocity as the river deepens.[9]

Imagine a river flowing through a forest. The geometry of the channel—whether it is wide and shallow or narrow and deep—dictates how much water it can carry. The slope of the riverbed controls how rapidly the river runs. In the meantime, the water may be slowed down by roughness of the riverbed influenced by rocks, branches, and debris, creating rapids or calm pools.

In the domain of open channels, the flow conditions can either remain constant over time (steady flow) or vary (unsteady flow). Steady flow is akin to a calm, predictable river, where the depth, velocity, and flow rate do not change with time at any given point. Here, flow conditions only depend on space, not time, and the discharge through any cross-section of the channel remains constant over time. No matter where in the channel they are observed, Eulerian or Lagrangian, the water level and flow rate are always the same. For water conveyance infrastructure like canals and aqueducts, the constant discharge is a blessing for engineers and designers as it makes calculations and forecasts easier. It makes it possible to take a straightforward approach to design, with a focus on ensuring that hydraulic structures can accommodate the required discharge under controlled velocity conditions.

Time-varying motion occurs when the flow characteristics (pressure, depth, and velocity) at any location in the channel change over time. This phenomenon is also referred to as unsteady flow or transient flow. It is the wild card of water movement, typical of urban stormwaters, river floods, or seasonal snowmelts. Because time-varying flow is dynamic, it is more difficult to investigate, model, and predict flow patterns. For instance, the fluctuations in water levels and flow rates make it difficult to forecast and control flood waves. Engineers need to take this variability into consideration when designing flexible systems that can withstand a variety of situations, from droughts to deluges, while also preserving the vital role of unsteadiness in shaping ecosystems and landscapes.

Transitioning from the temporal domain to the spatial domain, open channel flows can also be classified according to how their features change along

[9] Leopold, L. B. (1997). *Water, rivers and creeks*. Sausalito: University Science Book.

the length of the channel. Uniform flow maintains a constant depth and velocity throughout its path, representing an ideal state where the channel's slope, shape, and roughness are perfectly balanced with the flow's volume and speed. Uniform flow is pivotal for designing long stretches of canals or drainage ditches, where maintaining a consistent flow is essential for efficiency and effectiveness.

On the other hand, when flow depth and velocity vary along the channel's length, non-uniform flow occurs. This can take the form of gradually varied flow, where changes occur over a long distance, or rapidly varied flow, characterized by abrupt changes over a short distance. Gradually varied flow might be observed as water moves from a steep slope to a flatter area, requiring careful management to prevent erosion or sediment deposition. Water abruptly accelerates or decelerates near structures such as spillways or weirs, resulting in a rapidly changing flow. This state calls for a design that can accommodate these dynamic conditions. These fluctuations, whether they happen gradually or rapidly, challenge engineers to accommodate these changing moods in their design (Fig. 3.11).

The flow regime is determined by the balance between gravitational force and flow inertia. Another dimensionless number is useful here, as it is crucial for differentiating between various flow patterns. The Froude number, $Fr = V/\sqrt{(gy)}$, with y representing flow depth and V streamflow velocity, is a dimensionless quantity that relates the flow's inertia to the gravitational force.

Open channel flow can follow three different regimes: subcritical, critical, and supercritical. Subcritical flow, characterized by a Froude number smaller than unity, is tranquil and dominated by gravity. In this regime, any disturbance upstream can influence downstream conditions. It is useful when control and predictability are desired, such as building river regulation facilities or irrigation canals (Fig. 3.12).

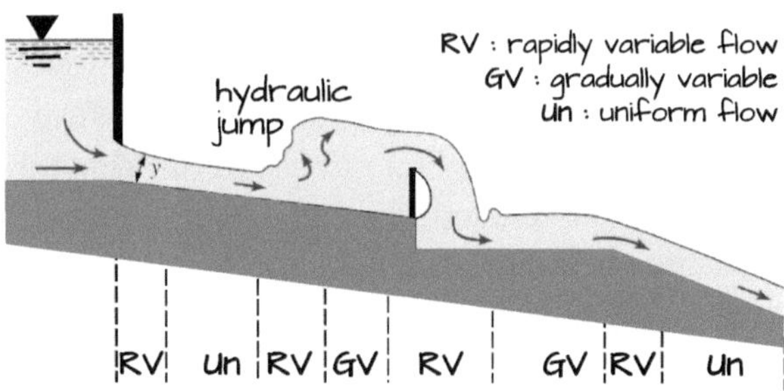

Fig. 3.11 Steady flow states

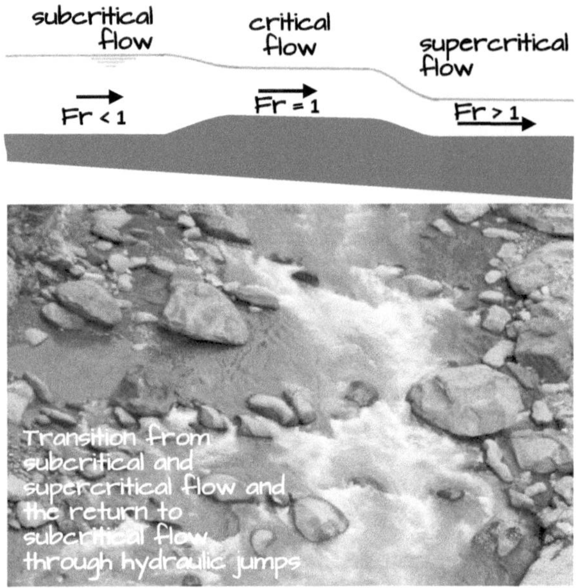

Fig. 3.12 Flow transitions

The flow reaches a critical condition when its velocity equals the wave velocity generated by a disturbance or obstruction. Under these conditions, Froude number of exactly one and specific energy is minimum. Critical flow conditions signify a balance between overcoming gravitational force and overcoming frictional resistance. This state is sensitive to even the smallest changes in the channel. In order to build water control structures that maximize flow conveyance while avoiding instability, engineers sometimes employ critical flow concepts.

Supercritical flow has a Froude number greater than unity. This means that water moves faster than wave velocity generated by a disturbance. This regime is marked by its sensitivity to variations in the geometry or slope of the channel, which can cause phenomena such as hydraulic jumps when transitioning back to subcritical flow. An appraisal of supercritical flow is essential for designing high-velocity canals, weirs, and spillways where managing the energy and momentum of is crucial for safety and functionality.

Each one of these flow characteristics and regimes plays a vital role in the design and management of water resources. By grasping the fundamentals of steady versus unsteady flow, uniform versus non-uniform flow, and the various flow regimes, engineers and environmental scientists can better predict, manage, and protect this invaluable resource.

As we delve deeper into the realm of open channel flow, we come across ideas that are not only fundamental but also revolutionary, providing new perspectives that close the knowledge gap between water engineering and theoretical hydrodynamics. Critical depth and hydraulic jump, together with the fundamentals of energy and momentum conservation, are two important ideas in this field.

In open channel flow, critical depth denotes a transitional point that separates subcritical and supercritical flow regimes. This pivotal concept describes the state of minimum energy for a given flow rate, where the flow velocity equals that of a wave caused by any perturbation of the flow. In applications, the critical depth must be accounted for in designing channels and structures that will function properly and safely under a range of flow circumstances.

On the contrary, the hydraulic jump occurs when a high-velocity, supercritical flow transitions to a lower velocity, subcritical flow, resulting in a quick rise in the water surface. There is a significant energy dissipation during this process. Its pattern can be very different depending on the properties of the upstream and downstream flows. These patterns vary from a standing wave to a stable or intermittent jump, depending on the upstream Froude number and the relative depth of upstream and downstream flows (Fig. 3.13). In order to control flow transitions in hydraulic structures, reduce the erosive potential of supercritical flows, and enable safer passage through channels and over spillways, it is essential to control the hydraulic jump.

Open channel flow analysis directly benefits from the fundamental principles of momentum and energy conservation. As mentioned above, the total mechanical energy of the water is the sum of its kinetic energy (due to motion), potential energy (due to elevation), and pressure energy (Fig. 3.14). When considering losses from friction and other resistances, conservation of energy suggests that the total energy of the flow is conserved along the flow path. This principle is applied in the analysis of flow profiles in rivers and canals, particularly in the construction of canals and associated hydraulic structures to guarantee effective flow conveyance.

The principle of conservation of momentum is equally important, highlighting the flow's ability to withstand modifications in motion. This idea is especially helpful in scenarios where there are abrupt changes in the wetted area or direction of streamflow, such as expansions, contractions, and bends in rivers and canals. It aids in the design of control structures and transitions, ensuring that they can withstand the forces applied by the flowing water and continue to operate as intended under a range of flow conditions.

Understanding rivers and efficiently designing, operating, and managing canals and other water systems depends on accurate measurements and

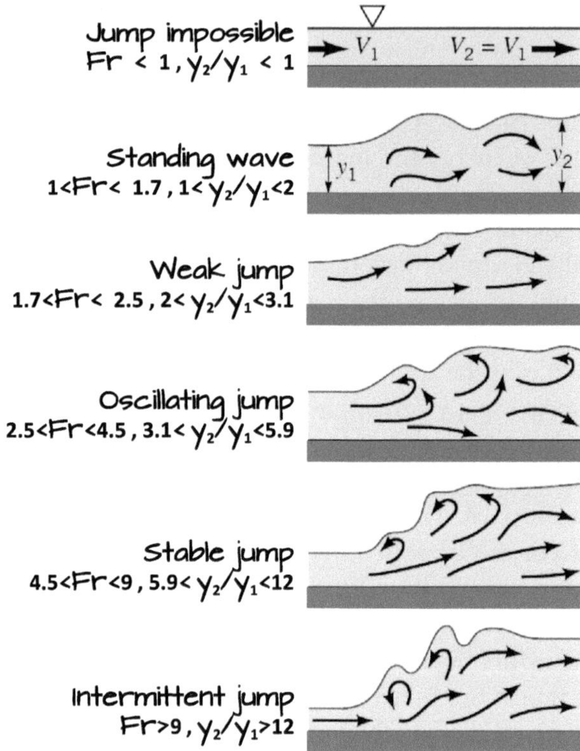

Fig. 3.13 Hydraulic jump

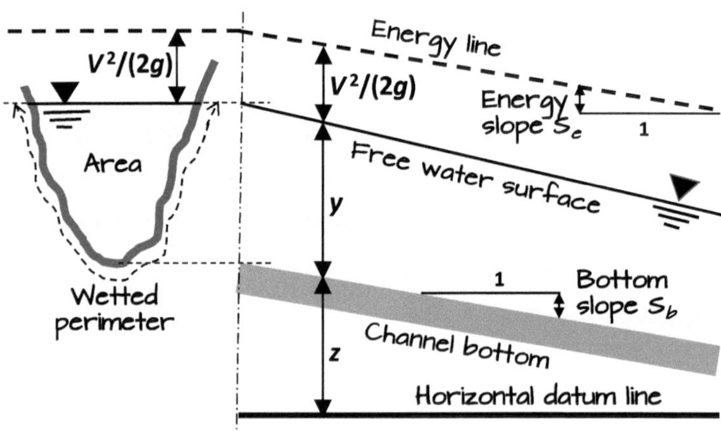

Fig. 3.14 Energy share of open channel flow

analysis of flow rates and velocities in open channels. To do this, a variety of tools and techniques are used, from basic mechanical instruments such as current meters to complex electronic sensors, including remote sensing from space. In situations where it is not possible to measure velocity directly, estimates are obtained indirectly by measuring the wetted cross-sectional area, and estimating the average depth of streamflow, y (Fig. 3.2).

The Chezy formula is a commonly used method for assessing flow parameters and estimating flow rates in open channels. It is a straightforward, yet powerful, method of examining how the geometry of the channel, particularly its roughness, and the slope can affect the speed of water. The formula is expressed as $V = C\sqrt{(RS)}$ where C is a friction coefficient and S the water surface slope. The cross-sectional area of flow divided by the wetted perimeter yields the hydraulic radius R, which for large channels equals the average water depth y. This pioneering formula was developed in 1768 by French physicist and engineer Antoine de Chézy while designing Paris's water canal system.

Since the Chezy coefficient C varies with the hydraulic radius, further empirical tools are used in practice, such as the Manning equation given by $V = R^{2/3}S^{1/2}/n$, with n denoting a friction factor depending on the channel's roughness.[10] Irish engineer Robert Manning developed this extension and modification of the Chezy approach in 1889, while French civil engineer Philippe Gaspard Gauckler pioneered it in 1868. The Swiss water engineer Albert Strickler refined the Gauckler approach in 1923. In practice, the Gauckler-Strickler roughness coefficient is the reciprocal of Manning's n.

In rivers, the most important factors that affect the selection of appropriate channel n values is the type and size of the materials that compose its bed and banks. A boulder bed river's friction factor can be up to three times higher than that of a sand bed river and twice that of a gravel bed river (Fig. 3.15). This indicates that streamflow in a sandy river is affected by half of the friction in a gravel bed river, which is a half of that of a boulder bed river. Other factors influencing channel roughness are the degree of irregularity, variation in channel cross-sections, effect of obstructions, amount of vegetation, and the degree of meandering.[11]

[10] Take note that $R^{2/3}$ means the cubic root of the square hydraulic radius R, and $S^{1/2}$ is the square root of slope S.

[11] See, e.g., Arcement, G. J. Jr., & Schneider, V. R. (1989). Guide for selecting Manning's roughness coefficients for natural channels and flood plains, U.S. Geological Survey Water-Supply Paper 2339; and National Academies of Sciences, Engineering, and Medicine (2024) Selection and Application of Manning s Roughness Values in Two-Dimensional Hydraulic Models, Washington, DC: The National Academies Press.

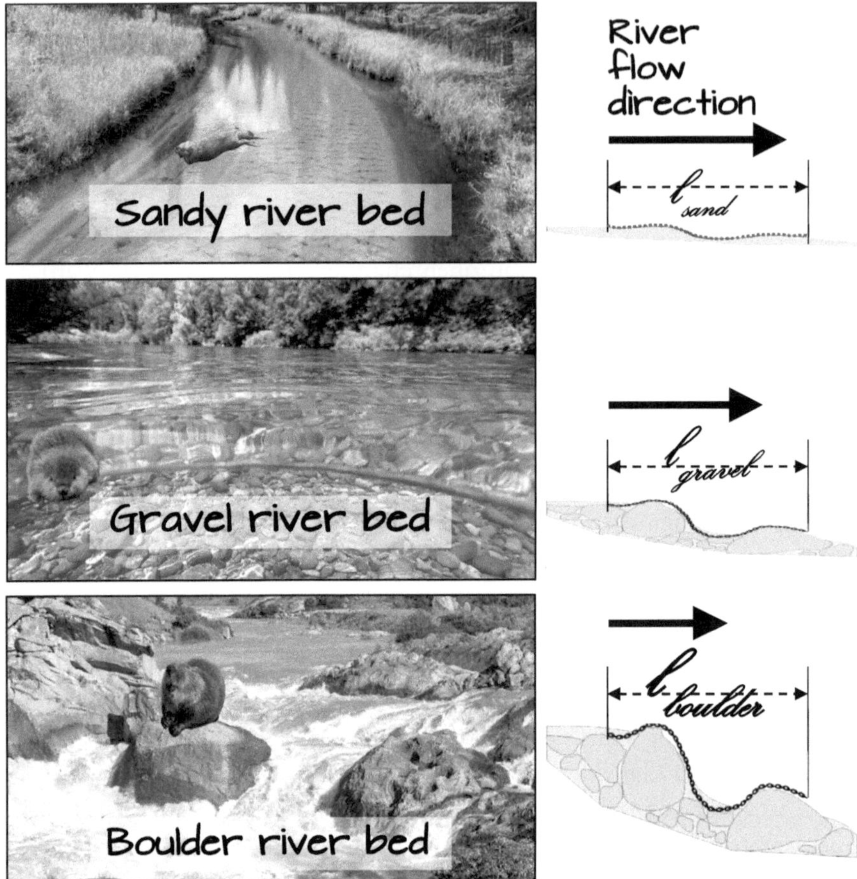

Fig. 3.15 Friction

Water conveyance and control structure are designed using open channel flow principles. It is essential to comprehend flow dynamics to properly design systems that are not only effective and efficient but also sustainable and environmentally friendly. These include flood control techniques, irrigation canals and aqueducts, and drainage systems in both urban and rural areas, among others.

Water Moves Everywhere

Water runs through all of nature, not just in rivers, creeks, and artificial waterways like aqueducts and canals. Every land mass on the planet experiences groundwater flow, infiltration, and percolation. In the oceans, there are sea currents, waves, tides, and seiches. Atmospheric water undergoes complex phase transitions between the vapor, solid, and liquid states when precipitation occurs.

Freshwater comes mostly from groundwater, with the exception of glaciers and permanent ice sheets. Infiltration is the process by which water enters the soil from the surface, while percolation refers to water moving vertically and horizontally through the soil layers once it has infiltrated. The way that air is trapped in the pores of the aquifer and interacts with water makes these processes extremely complex.

Understanding groundwater flow in a saturated aquifer is made considerably easier by Darcy's Law, which provide an estimate of the rate at which water will flow through a porous media as driven by the difference of pressure or hydraulic head. In practice, the average flow velocity U is proportional to the slope s of the surface of the groundwater table. The constant of proportionality K, which reflects how easily water can flow through the pores of terrain, is called the medium's hydraulic conductivity: $U = Ks$ (Fig. 3.16). This law was discovered by Henry Darcy in 1855 while conducting column experiments in a hospital at Dijon, France, to determine "the laws of water flow through sand." His findings indicated that, "for sand of comparable nature, one can assume that the discharge volume is directly proportional to the head and inversely proportional to the thickness of the layer traversed."

Somewhat similar to the Manning's friction factor, hydraulic conductivity, measured in SI units of meters per second, mainly depends on the grain size of a porous medium. Other factors include grain shape and sorting, soil texture, and water temperature. For example, the hydraulic conductivities of

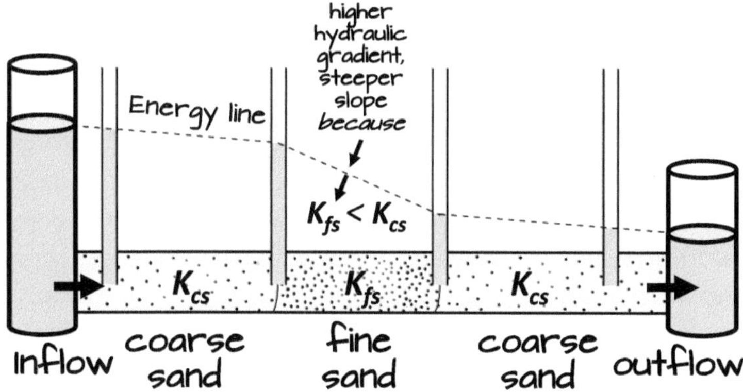

Fig. 3.16 Darcy's law

fractured rocks are ten thousand times higher than those of sandstone, and the K value for sandy soil may be one thousand times higher than that of silt.

Currents are the horizontal movements of water in oceans, rivers, and other water bodies. They are influenced by a multitude of factors, such as temperature, salinity, wind, and tides. Different types of currents, like surface, deep, and tidal currents, play a vital role in marine life, distributing heat and spreading nutrients.

Waves are disturbances that generate through water, transferring energy without significant mass transport. Ocean waves are a common example. These rhythmic undulations on the sea's surface are due to the interaction of wind with the water. Picture yourself standing on the shore, watching the sea. As the wind blows across the surface of the water, its energy is transferred to the water particles, causing them to move in a circular pattern. Waves are produced by this motion and propagate across the ocean's surface.

Waves vary in size, shape, and strength depending on factors like wind speed, duration, and the distance over which it blows, sometimes called "wind fetch." The strength and duration of the wind force determines the waves' height and the wavelength, which is the distance between them. Waves transfer energy from one place to another over thousands of kilometers in the sea.

As waves approach the coast, their properties change due to their interaction with the seafloor. Because the circular motion occurring in deep water collapses into more complex trajectories, they might get steeper and higher until they crash upon the shore as surf (Fig. 3.17). Waves play a crucial role in shaping coastlines, transporting sediment, and influencing marine ecosystems.

Sea levels increasing and falling due to tides is the result of the combined gravitational forces of the Moon and the Sun as well as the effect of the Earth and Moon orbiting each other (Fig. 3.18). The tidal motion varies on timescales ranging from hours to years due to a number of factors, and are the largest source of short and medium-term sea-level fluctuations. In addition, sea levels are subject to change because of thermal expansion, wind forcing, and barometric pressure changes, resulting in storm surges, especially in shallow seas and near coast lines.

Sea seiches are standing waves are mostly caused by variations in barometric pressure. They can be seen in ocean bays in Canada, Japan, and Sri Lanka, as well as in seas like the Baltic and Adriatic. They can even flood Saint Petersburg, which is built on a former marshland. Autumn overflowing of the Neva River is frequently caused by seiches as a consequence of a low-pressure area in the North Atlantic creating cyclonic lows in the landlocked Baltic Sea. The term seiche was initially used in the alpine region to describe the

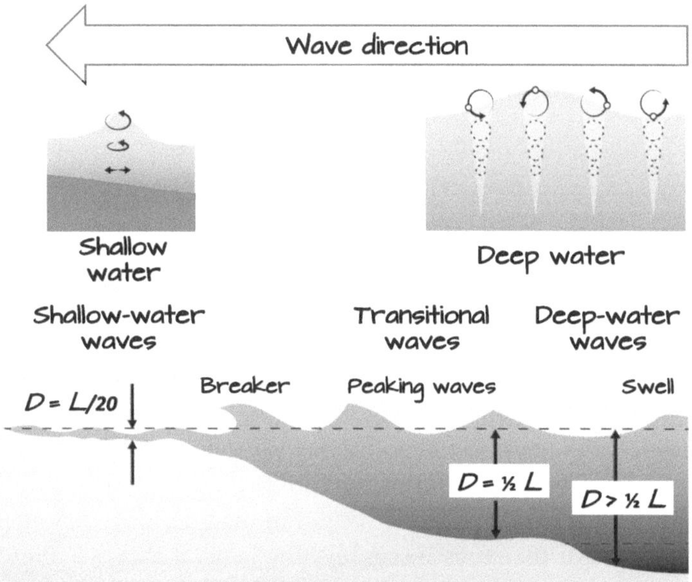

Fig. 3.17 Sea waves

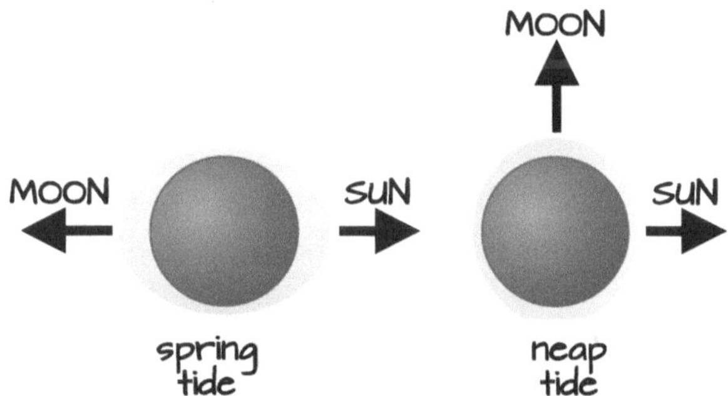

Fig. 3.18 Tidal attraction

oscillations in alpine lakes,[12] as first studied by Swiss hydrologist François-Alphonse Forel in 1890.

Other standing wave phenomena are observed in enclosed or partially enclosed water bodies, such as lakes, reservoirs, harbors, and caves. Seismic activity, long-period or infragravity waves, wind and air pressure changes are examples of geologic and meteorological drivers. Similar to a tidal wave, a tsunami is a breaking wave caused by the displacement of a large volume of

[12] This Swiss French dialect word "sec" or "sèche" comes from the Latin word "siccus" meaning "dry".

water, generally in an ocean or a very large lake. Earthquakes, landslides, volcanic eruptions, glacier calvings, meteorite impacts and other disturbances above or below water all have the potential to generate a tsunami.

Theory Versus Practice

Ever since the beginning of time, people have struggled to understand the laws governing water flow, their impacts, and how humans have devised solutions to harness and manage water. Applying the fundamentals of mass, energy, and momentum conservation allowed for the advancement of knowledge of water in motion, which had previously been based on holistic approaches and metaphysical conjectures. Throughout history, the development of knowledge concerning water motion has been a blend of practical needs, such as irrigation and water supply, human curiosity to understand the natural world, and the effort in establishing general theories. However, only practice claimed for field observation of nature.

> The test of science is its ability to predict.[13]

Scientific theories begin with conjectures—creative guesses or hypotheses about how the water world works. These are audacious and creative speculations that go beyond the direct realm of current observations and understanding. Scientists attempt to refute these conjectures through rigorous testing and experimentation. A theory may only be considered scientific if it is refutable by an observation or experiment, which implies that it must be falsifiable. According to Karl Popper, a theory is refutable (or falsifiable) if it can be logically contradicted by an empirical test.[14] This approach recognizes that we can never completely prove a theory correct; all we can do is try to refute it. When a notion remains unproven despite multiple and different attempts to refute it, we consider it tentatively accepted or validated, but not proven beyond a reasonable doubt.

According to Galileo's tidal theory, the Earth's rotation and revolution around the Sun are what cause the tide fluctuations. He thought that these motions worked together to accelerate and decelerate the Earth's waters, which in turn caused sea levels to rise and fall like water fluctuating in a moving

---

[13] Feynmann, R. (1963). *Lectures on physics*, Vol. 2, Ch. 41.

[14] Popper, K. (1934). *Logik der Forschung*. Berlin, Julius Springer, Hutchinson & Co (*The Logic of Scientific Discovery*, in German).

bowl. He introduced this theory in his *Treatise on the Tides*, written in 1616, and further detailed it sixteen years later in his masterpiece, the *Dialogue Concerning the Two Chief World System*. Here, he argued against the need for a lunar or solar gravitational influence on tides.

Solid knowledge clashed with Galileo's theory. The ancient Greeks were aware of the connection between the Moon and Sun and the tides, except Aristoteles who believed the tides were caused by winds driven by the Sun's heat. However, the "first of the moderns" who eventually developed the scientific method, Bernardino Telesio, followed Aristoteles in his masterpiece *De Rerum Natura Iuxta Propria Principii* (1570). According to Telesio, the general cause of the continuous motion of the sea is its simmering, produced by solar heat and the formation of vapors "which attempt to come out but are hindered in this attempt by the sea above; as a consequence they raise and swell it just like the spirit that fire generates in the water".[15]

Conversely, knowledge in China advanced to the point where the first tide-table for estimating the heights and times of tidal bores—waves that travel up rivers or narrow bays against the direction of the river's current—was constructed around 1000 CE. The twice-daily timing of tides and the monthly cycle of spring and neap tides are both influenced by the Moon, as Bede the Venerable made evident in his early middle ages treatise on tides, *De temporum ratione.*[16] Dante Alighieri remarks the Moon's influence on tides in his *Divine Comedy*. Simon Stevin in his *Theory of Ebb and Flood* (1608) dismissed a large number of misconceptions that still existed about ebbs and floods, pleading for the idea that the attraction of the Moon was responsible for tidal motion. However, the confutation of Galileo's theory was not easy at this time.

The falsification and eventual refutation of Galileo's theory on tides is not straightforward because it involves several key developments. First, one makes use of empirical data. More methodical and accurate observations of tides were performed over time, highlighting inconsistencies in Galileo's theory. It struggled to explain the regularity and timing of tidal patterns observed. For example, it did not accommodate the occurrence of two high tides and two low tides within around 24 h observed in many locations.

The advancements in gravitational theory provide another argument. The most significant challenge to Galileo's theory came from Newton's law of

[15] See: Omodeo, P. D. (2019). Telesio and the Renaissance Debates on Sea Tides. In: C. Lüthy (Ed.), *Bernardino Telesio and the Natural Sciences in the Renaissance*, Chap. 6:116–145, Leiden: Brill. This paper quotes: Bernardino Telesio, *De mari, in De iis que in aere fiunt et de terremotibus,* as translated by F. Martelli, a book edited by L. Franco, Cosenza: Bios, 1990.

[16] *The Reckoning of Time* is an English era treatise written in Medieval Latin by the Northumbrian monk Bede in 725.

universal gravitation (1687). Newton postulated that the gravitational force of the Sun and Moon on the Earth's seas caused the tides. The gravitational pull explanation could account for the observed regularities of the tidal patterns, including their timing and magnitude, far more effectively than Galileo's approach.

More recent prospects include mathematical models and empirical testing. The capacity to forecast tidal movements based on the gravitational pull of the Sun and Moon increased along with knowledge of gravity and its effects on celestial bodies. In contrast to Galileo's predictions, the gravitational theory of tides was increasingly confirmed by mathematical models and empirical data.

The refutation of Galileo's tidal theory took time to come about; it evolved as part of the broader shift toward the Newtonian paradigm in physics and astronomy. Only the accumulation of evidence supporting the gravitational theory of tides, combined with the successful application of Newton's laws to a wide range of astronomical phenomena, led to the gradual rejection of explanation by Galileo, one of the fathers of modern science.

4

The Water Cycle

Save the rain: each raindrop is a kiss from heaven.
Friedensreich Hundertwasser, Poster for the Norwegian Society for the Conservation of Nature, 1981

The River's Headwaters

Mankind has always perceived the river's headwaters as a mystical, magical, iconic place. It is a space of physical and political geography that, in reality, belongs first and foremost to spiritual geography. As Minerva to the Romans, Athena was the Greek goddess of wisdom, arts, and war. Her birthplace was the headwaters of the Triton River, which is a tributary of the same-named lake. In Ovid's *Metamorphoses*—the masterpiece by one of the three canonical poets of Latin literature—this location is mentioned as Tritonia. The river vanished and its exact location is still unknown; however, it was most likely in Libya.

The Germanic, Mesopotamian, and Eastern traditions, as well as Greek and Roman mythology, have all drawn inspiration from the holiness of water and the mystical significance of springs.[1] Water is often described in Celtic mythology as a spring, a source, a caring force, and the lifeblood of plants. The ability of flowing water to regenerate and create new life makes it the archetype of life itself.

---

[1] Seppilli, A. (1977). *Sacralità dell'acqua e sacrilegio dei ponti*. Palermo: Sellerio (*Sacredness of water and sacrilege of bridges*, in Italian).

R. Rosso, *Five Easy Pieces on Water*, https://doi.org/10.1007/978-3-031-69276-5_4

Men have been exploring the land surface since prehistoric times, looking for river headwaters. Whether driven by the need to learn more, lust for conquest, expand their economies, or satisfy their greedy curiosity, the world's rulers have frequently looked for the headwaters of major rivers. Between 62 and 67 AD, Emperor Nero set out on an expedition to locate the source of the African river that fed Egypt from the south to the north. After traveling up the Nile to equatorial Africa, three centurions came to a location that was surrounded by large, sometimes impenetrable wetlands and a river that flowed between rocks. It was maybe Lake No, which is located where the White Nile and the Bahr Al Ghazal rivers converge. Some academics contend that the expedition ventured considerably farther, arriving at Murchison Falls in northwest Uganda (Fig. 4.1).

The latter conjecture is supported by an outstanding Nero's advisor, Seneca, a Stoic philosopher, dramatist, statesman, and satirist. He located the waterfalls near Lake Albert, a dropping of about 50 meters in a branch of the White Nile, emissary of Lake Victoria:

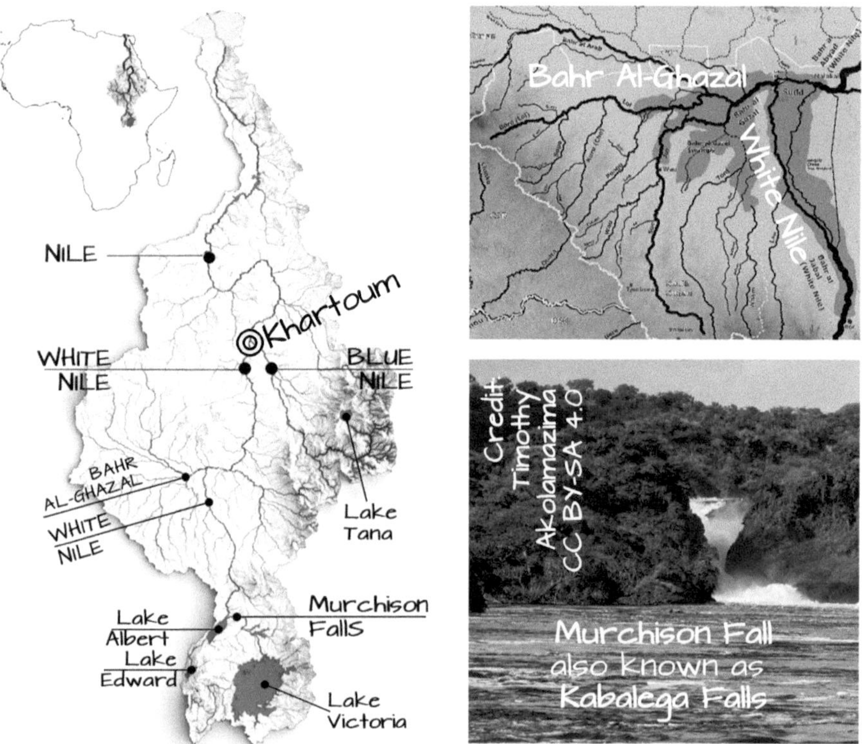

Fig. 4.1 The search for the Nile sources

The river gushed out with power from two rocks that we saw.[2]

A recurring storyline in contemporary literature is the eerie quest for the river's headwaters. From the start of Joseph Conrad's *Heart of Darkness* until the end, Charles Marlow—the sailor, co-protagonist, and storyteller—travels up the Congo River and experiences a number of physical and psychological transformations. His mission to find the headwaters serves as a metaphor for his existence as a modern-day Thor setting out to subdue Hvergelmir, the roaring, boiling spring that serves as the source of all rivers. In science fiction, the voyage to the river's headwaters represents humanity's greatest ambition. The science fiction author Philip Farmer narrates the tale of Samuel Clemens aka Mark Twain, who is designated to locate the river's headwaters, which are thought to be the home of Riverworld's creators.[3]

Several millennia of human history have passed before the headwaters of the Nile were discovered, a riddle that belonged to Herodotus, Ptolemy, and Nero. Gian Lorenzo Bernini was commissioned by Pope Innocent X in the seventeenth century to design the *Fountain of the Four Rivers,* which symbolizes the Danube, Ganges, Rio de la Plata, and Nile, the four major rivers of the four continents (Fig. 4.2). Because its headwaters had not yet been found, Bernini—the architect who pioneered the Baroque style in sculpture—depicted the Nile as a giant with a cloth covering his face to represent the darkness about its unknown sources. Wilbur Smith's latest book, *The Quest,* shows how the hunt for the Nile's headwaters continues to captivate and astound people even after the mystery was solved in the middle of the nineteenth century.

The Mystery of Springs

While finding a river's headwaters is a long-standing, lofty human desire, people have also been chasing a different mystery for a very long time. From what source does a spring's water come? Nowadays, a primary school student could immediately answer: "From the rain that has infiltrated the mountain." Before getting this seemingly simple solution, humanity spent ages or possibly millennia working, discussing, conjecturing, and refuting more or less fanciful theories. At times, disagreements would turn violent, forcing the Holy Inquisition tribunals to arbitrate the conflicts.

---

[2] Seneca (65) De Terrae Motu, in: *Naturales Quaestiones*, Liber VI, 8,5.

[3] Farmer, P. J. (1971). *The fabulous riverboat.* New York: G.P. Putnam's Sons.

Fig. 4.2 Fountain of the four rivers in Rome

According to a biblical verse, it is possible that the ancients guessed the answer.

> All streams run to the sea, but the sea is not full; to the place where the streams flow, there they flow again.[4]

It was more about accepting the worthlessness of human things than it was about the actual item. It was never interpreted literally. For a very long time, people have believed that a spring's water originates from the dark Tartarus, which is at the bottom of the abyss that descends beneath the ground.[5]

Thales, who lived in the sixth century BC, proposed that there is just one ultimate substance, water, on which everything of nature depends. He went on to explain that rivers are made possible by the fact that the Earth floats on water. Other natural philosophers, who took the question more seriously than Thales, led the way in modern knowledge. In the fifth century BC, Anaxagoras

[4] *Ecclesiastes* (1:7).

[5] *The Iliad* (Book VIII, 368–9).

thought that the sun's heat carried seawater into the atmosphere, from whence it descended as rain. According to him, rainfall that seeps through the earth fills huge subterranean reservoirs, which then supply water to the rivers. Theophrastus of Ephesus, a century after this reasonable intuition, acquired a deeper understanding of the atmospheric phase of the water cycle and explained how water vapor condensation and cooling that rises from the sea and is carried by winds results in precipitation.

The Basilica of Fano is the sole notable work by Vitruvius, who lived two thousand years ago. Despite this, he is the most well-known Roman architect. Once he had a comfortable pension, he concentrated on authoring the treatise *De Architectura*, while Emperor Augustus began an amazing building spree. He aimed to standardize construction methods for buildings, arenas, fortifications, roads, and aqueducts, wondering where the water came from because it is always necessary but occasionally annoying. He deduced that groundwater was fed by rains and snow infiltrating the ground, and the headwaters of the rivers prove it.

> The headwaters of the rivers prove it, as indicated on the maps or described, and they arise the most and are the largest from the north. Here first in India, the Ganges and the Indus start from Mount Caucasus; in Syria, the Tigris and the Euphrates; in Asia, and especially in Pontus, the Boristhenes, the Ipani, the Tanai; in the Colchis, the Phasis; in Gaul, the Rhone; in Belgica, the Rhine; across the Alps, the Timavo and the Po; in Italy, the Tiber. In the Maurusia, or Mauritania, from Mount Atlas, the melting waves, swollen and heavy, spread over the Earth.[6]

Muhammad Al-Karaji, a Persian mathematician and engineer from Baghdad, shared Vitruvius' conjecture ten centuries later. His wonderful treatise on the extraction of hidden waters states that springs are formed from water flowing in the subsoil, while water on the surface of the earth is caused by excess rain and snow. The soil is subjected to percolation by the rain and snow that has melted, and the excess water is released into the sea as runoff. Rain and snowmelt percolate into the soil, and runoff releases the extra water into the ocean.

Al-Karaji conjectures were driven by an ancient heritage: the qanats, which are tunnels that transport deep water from the mountain interior to the plain below (Fig. 4.3). Most of the water used for drinking and irrigation in the warm, desert regions of Persia came from these tunnels, underground

[6] Galiani, B. (1790). *The architecture of Marco Vitruvio Pollione*, Book VII, Chap. II, Naples: Fratelli Terres (in Italian).

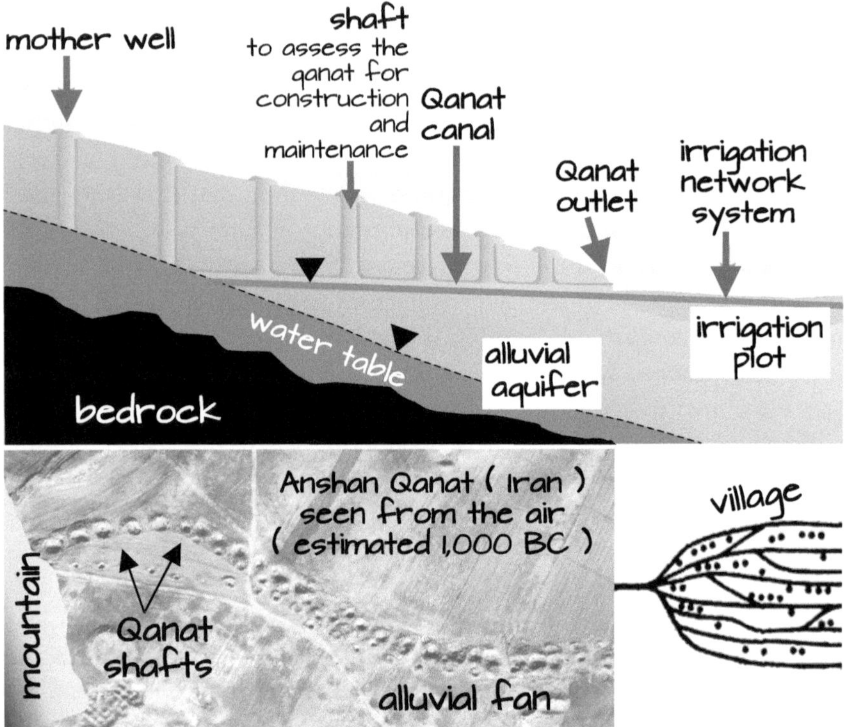

Fig. 4.3 Qanat

aqueducts that prevents significant water evaporation loss during long-distance water transportation in hot, dry conditions. The concept of qanats reached the East, India, and China, as well as the West, Mesopotamia, and Northern Africa. The idea that river water originates in the atmosphere was quickly accepted in China and India, but the West was slow to adopt these theories.

The water cycle is a basic concept in scientific culture today. Every elementary school student has created a color sketch depicting the continuous water exchange between the Earth and the sky, which is brought about by the gravity and heat of the Sun (Fig. 4.4). The water cycle's response is a major concern when it comes to the climate effects of global warming.

Water evaporates from wet soil, from growing plants' leaves, from natural and artificial lakes and from the oceans which are the major source of atmospheric water. It is transported by air as a gas called water vapor. Air water vapor condenses when it changes its state from a gas to a liquid or solid state, at which point gravity causes it to fall as rain or snow over the land and

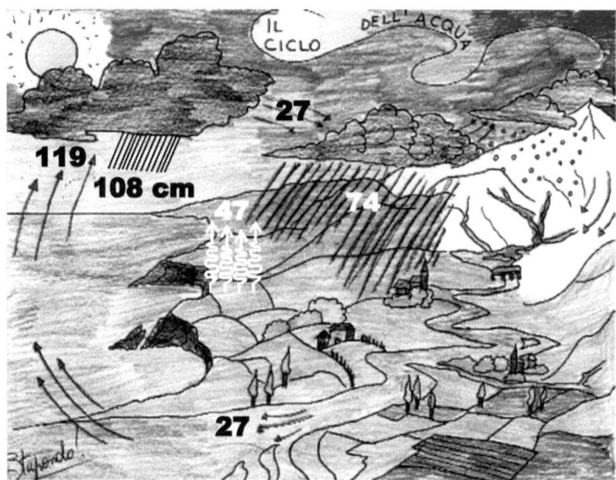

Fig. 4.4 The hydrologic cycle

oceans. Precipitation feeds lakes, rivers, and oceans, which ultimately receive it from the rivers. Water is constantly exchanged back into the atmosphere through evaporation from land surfaces and mainly oceans. Water flows continuously from the Earth into the atmosphere and back again.

This process has long been disregarded, misunderstood, or misinterpreted, even though it seems elemental now. Mass conservation, evaporation, condensation, and infiltration were not understood by scientists and philosophers until the second half of the seventeenth century. Even a notion as clear as the hydrological cycle would be impossible to comprehend without a solid explanation for the springs' origin. Even with all of the wisdom that antiquity has left us about water, the most fundamental issue was overlooked: how springs originate. Only in the late modern era was the hydrological cycle a part of Western countries' cultural heritage. For a long time Western scholars ignored Leonardo da Vinci's aphorism maintaining that, in rivers,

> the water that you touch is the last of what has passed and the first of that which comes; so with present time.[7]

Ultimately, these scholars were justified, because Leonardo did not draw any direct, clear, and definitive conclusions from his statement.

[7] Leonardo da Vinci, *Codex Trivulzianus*, 34v.

A Never-Ending Story

Water is being continuously exchanged between the Earth's crust and the atmosphere. The force of gravity and the heat from the sun cause this exchange to occur. Wet ground, growing plant leaves, rivers, lakes, and reservoirs all supply water evaporation. Water vapor travels through the air as a gas, then it condenses and transforms from a gas to a liquid when it cools, falling as rain or snow. Precipitation is what feeds the rivers, lakes, and oceans. Water from rivers flows into the ocean. Water flows in a circular motion from the atmosphere to the earth and back again.

This continuous process is known as the "hydrological cycle" with "cycle" focusing on dynamics as well as stability, periodicity, and repeatability. The basic concept is that water moves globally within the Earth's closed system, changing its state from solid to liquid to gas. Each drop of water follows a trajectory that leads it from the air, ground, land, or sea. Occasionally, this path takes it as runoff across the land surface or deep into the Earth's crust. The energy required for motion is provided by the Sun and gravity.

A look at the Earth from space reveals an aquatic world. The domination of the Earth's surface by the oceans shows a mostly blue planet (Fig. 4.5a). The Earth's surface is blue due to more than 1.34 billion cubic kilometers (cukm) of water and ice. However, the volume of the Earth's water is minuscule in comparison with the planet's solid mass. Despite the appearance of the opposite on the surface, as discovered by Copernicus in the sixteenth century, there is more solid matter than water on Earth. The Earth's volume is just over one trillion cukm, while the hydrosphere contains only 1.39 billion cukm of water and another eight million are believed to be stored in the Earth's mantle, most

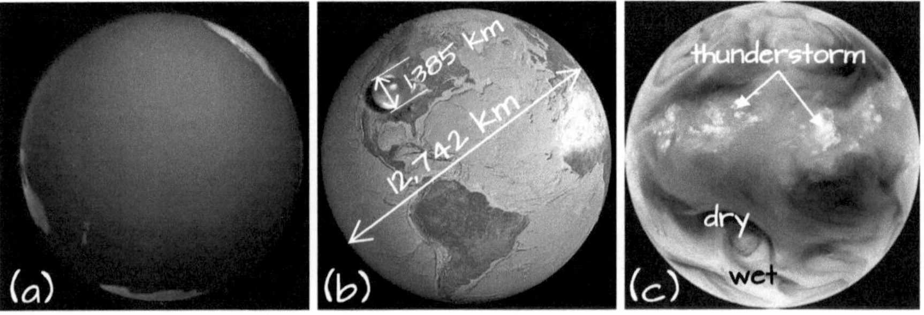

Fig. 4.5 Earth's water: the blue planet (a), volumes of water and planet (b), uneven water vapour distribution (c)

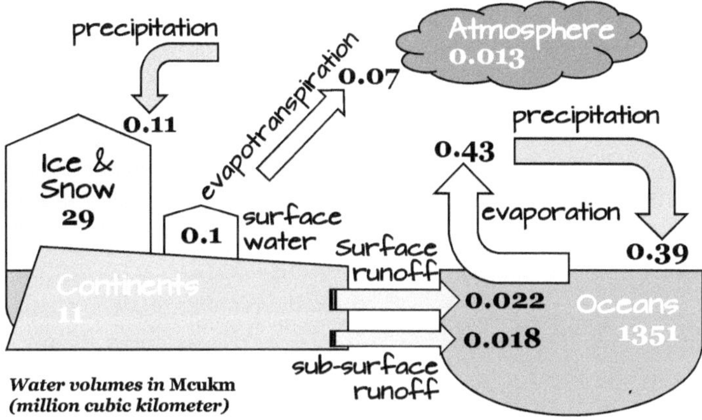

Fig. 4.6 Earth's water distribution

at depths between 400 and 650 km. Thus, our water world is less than 0.2% of planet Earth in volume[8] (Fig. 4.5b).

Water can be fresh or salty. The hydrosphere contains 1.35 billion cukm of salt water, that is about 97% of the Earth's water (Fig. 4.6). The oceans hold the majority of salt water, while salt aquifers and lakes contain just under one percent. The majority of solid water is contained in the glaciers of Antarctica, Greenland, and the highest mountain ranges; and a very minor portion is trapped in the permafrost. The total amount of freshwater is about 11.1 million cukm, less than one percent of the Earth's water. Groundwater accounts for 11 million cukm, while surface water accounts for only 0.007% or 70 per million of the hydrosphere. The atmosphere contains a small amount of water in gaseous form, roughly 13 thousand cukm.[9] Freshwater and atmosphere water vapor account for a negligible share of the available water in the world.

The first step in approaching the water cycle deals with completing a basic balance sheet, which creates a list of inputs or inflows, exits or outflows, and assets or water storage. Nature understands that these quantities need to be balanced and dislikes debits. The budget for water is comparable to that of a household. Still, compared to what a homeowner typically knows, the knowledge of water is far less trustworthy due to uncertainty and a lack of data.

It is difficult to develop an accurate hydrological balance for a stream, lake, or aquifer that supplies a city on a daily basis without accurate, detailed, efficient monitoring. When the hydrological system is a nation, a continent, or

[8] The volume of Earth with a volumetric diameter of 12,742 km is about one trillion cukm (1.082×10^{12} cukm).

[9] Kotwicki, V. (2009). Water balance of Earth/Bilan hydrologique de la Terre. *Hydrological Sciences Journal*, 54:5, 829–840.

the planet Earth, the uncertainty grows with decreasing time scale of analysis. While there are significant uncertainties when looking at a short period of time, the available information is generally adequate to estimate the annual budget of the municipal water supply under exam, likewise the annual balance sheet of a company. Here, the "company" is the aquifer or river that supplies drinking water, but it can be a larger system such as a drainage basin, a country, a continent, or the entire planet Earth.

The water budget of a hydrological system needs to account for the water that is found in the atmosphere, soil, subsoil, rivers and lakes, glaciers and ice sheets, and oceans. The inflows, or receipts on the balance sheet, come from direct flows to the Earth's crust. The exits are provided the Earth's crust's outflows. These fluxes can be expressed in a variety of units, including liters, cubic meters, hectoliters, or cubic kilometers—the unconventional unit of measurement that we used on a planetary scale. For the annual budget, each of these amounts is given as a volume per unit of time; one year for the annual budget.

We can present volumetric measurements in cubic meters or kilometers, as shown earlier, but it is more straightforward to grasp them when given in terms of depth, such as centimeters or millimeters, over the area under investigation, whether it is an irrigation plot, a basin, a country, or even the entire planet Earth. One centimeter of water depth would represent the amount needed to cover the study area to a depth of one centimeter. For example, one centimeter of water covering the United States would be equivalent to 78 billion cubic meters, roughly double the storage capacity of Lake Mead, where the Colorado River is impounded by the Hoover Dam. Similarly, one centimeter of rainfall over Italy would equate to approximately 30 million cubic meters.

In large-scale systems, water inflows considered as credit items are solely attributed to rainfall and snowfall. On average, Italy receives approximately 97 centimeters per year (cm/y), while the contiguous United States receives 68 cm/y, and the entire planet Earth receives around 100 cm/y. However, the spatial distribution of this water supply is highly uneven, and this disparity intensifies with the size of the area under examination.

There are several debit items that contribute to the withdrawal of water from the system under exam. On the global scale, these include streamflow, the seepage of groundwater into oceans, transpiration from plants, and evaporation from various sources such as lakes, ponds, swamps, rivers, and moist soil. Italian rivers discharge approximately 50 cm/y of water into the sea, which accounts for roughly 52% of the total precipitation inflow. Evapotranspiration, the combined process of evaporation and plant transpiration, makes up about 40% of the water loss, while the remaining 8% is lost through deep seepage. In the United States, rivers discharge about 23 cm/y of water into the sea annually, with the Mississippi River alone accounting for approximately 40% of the total.

Although the yearly precipitation in the basin exceeds the national average by nearly 10%, the Po River—encompassing 24% of Italy's total area—contributes only 16% to the annual water outflow. The extensive utilization of freshwater in agriculture within the Po Valley results in increased evapotranspiration rates, contributing to its limited hydrological yield. The influence of irrigation is noteworthy, given that a substantial portion of agricultural water is directly returned to the atmosphere.

The distribution of precipitation varies significantly across different locations. A substantial amount of atmospheric water contributes to oceanic systems compared to terrestrial areas: oceanic precipitation averages around 108 cm/year, whereas only 74 cm/year falls on land. Less than half (43%) of land precipitation reaches the oceans, with the remaining 57% returning to the atmosphere through evapotranspiration. Oceanic evaporation is approximately 119 cm/year, considerably exceeding sea precipitation rates.

Weather systems carry around the globe water vapor, and with it energy. As shown in the satellite image of Fig. 4.5c the distribution of water vapor over Africa and the Atlantic Ocean is uneven. Clear areas have high concentrations of water vapor, while dark regions are relatively dry. Thick thunderclouds are the brightest white areas.[10]

The Earth's fragile atmosphere, which brushes against both the vast oceans and sprawling continents, serves as a conduit that links the land and sea in the water cycle. Despite its vastness, the amount of water held within the atmosphere is relatively small, totaling just 12.9 thousand cubic kilometers. To put this in perspective, if spread evenly, it would form a layer about 2.5 cm thick over the entire surface of the Earth—roughly equivalent to one inch. When compared to the massive volume of precipitation each year, which amounts to 100 cm and totals 515 thousand cukm, the atmospheric water content appears minuscule. Consequently, water typically spends about 9 days in the atmosphere before being deposited back onto the Earth's surface. Recent data indicate that the global average residence time is approximately 8.9 ± 0.4 days, with an uncertainty range reflecting one standard deviation.[11]

---

[10] The image of Fig. 4.4.*c* was acquired on the morning of September 2, 2010 by SEVIRI aboard METEOSAT-9 (see: https://gpm.nasa.gov/education/articles/earth-observatory-water-cycle-overview)

[11] van der Ent, R. J., & Tuinenburg, O. A. (2017). The residence time of water in the atmosphere revisited. *Hydrology and Earth System Sciences*, *21*, 779–790.

The time water spends in the atmosphere changes depending on where and when you look. Factors like regional patterns, seasons, and large-scale weather systems affect evaporation, rainfall, long-distance water movement, and air mixing. Over the ocean, water tends to stay in the atmosphere for about two days less than over land. During winter in the Northern Hemisphere, the age of moisture in the air is typically lower compared to summer.

Scientific theories suggest, and observations support, the idea that in most parts of the world (though not universally), human-driven climate change is causing the atmosphere to hold more moisture. This increase happens faster than the rates of evaporation and precipitation are speeding up.[12] As a result, the chance of severe storms to occur may increase.

The chance of water evaporating from a certain place and then falling back as rain in the same spot is relatively low, especially in larger or open regions. This occurs because the movement of water vapor in the atmosphere tends to spread it out across broad areas. Pinpointing the exact amount of water that evaporates and eventually returns as rainfall to the original location is a complex task. It requires sophisticated climatological models that factor in local evaporation rates, atmospheric circulation and precipitation patterns.[13]

Generally, the recycling of precipitation is linked to subtle moisture transport and high rates of evaporation. In enclosed areas such as some inland seas or lakes, a greater portion of the evaporated water may indeed return as precipitation to the same vicinity. However, this phenomenon does not hold true universally for all places on Earth.

Based on the assumption that future trends will mirror those of the past, long-term statistical forecasts offer strong estimates of the average or expected values of hydrological factors like rainfall and river flow. These predictions also encompass the anticipated range of variation, as indicated by second and third order statistical measures. Using dependable extreme value statistical models, it is possible to forecast the likelihood of significant events such as severe droughts and floods. The reliability of these predictions improves with the length of time covered by the available data.

Long-term statistical predictions outperform short- and medium-term forecasts. Describing the water cycle within a mathematical framework for climate analysis and predicting future scenarios is simpler than generating mid-range water flow forecasts for weather prediction. Hence, there is a saying

[12] Gimeno, L. and 8 others. (2021). The residence time of water vapour in the atmosphere. *Nature Reviews Earth & Environment*, *2*, 558–569.

[13] Bisselink, B., & Dolman, A. J. (2008). Precipitation RECYCLING: Moisture sources over Europe using ERA-40 data. *Journal of Hydrometeorology*, *9*, 1073–1083.

sometimes attributed to Mark Twain: "climate is what we expect, weather is what we get."

The Water Balance

Let us delve deeper into the concept of cycle. The hydrological system is commonly understood to be an open system that is guided by the mass balance equation, even though it is only considered closed when viewed at the global scale. According to the principle of conservation for mass, the mass balance equation dictates that the change in system volume (*dS*) must balance any difference between the input (*In*) and output (*Out*) flows over a specified period of time.

$$dS / dt = In - Out.$$

We typically think of volume and mass as the same thing because liquid water is an incompressible fluid. However, mass is used in place of volume when considering different phases of water, such as ice or solid water. The solid components can also be converted to correspond to their equivalent in water, which is often accomplished using millimeters or centimeters of water depth.

The mass balance equation holds for a specific area or control volume, which must be clearly defined. Within this context, *In* and *Out* represent the flows crossing the boundaries of the system, such as the watershed of a drainage basin, but also the boundaries of a region, or even a continent.

We can divide the water balance into two parts by looking at surface and underground processes individually (Fig. 4.7). Changes in surface water storage dS_S occur on the ground surface for a variety of factors (Figure). These include positive inflows such as precipitation P, surface inflows Q_{IN} from upstream water courses, and subsurface returns Q_S from groundwater table. On the other hand, negative components include soil infiltration I, plant transpiration T_S, evaporation E_S, groundwater recharge Q_G, and river outflow Q_{OUT}:

$$dS_S = P + Q_{IN} + Q_S - Q_G - Q_{OUT} - E_S - T_S - I.$$

Three components of inflows and four components of outflows balance the change in volume dS_G of subsurface water. Inflows include infiltration I, groundwater recharge Q_G, and underground inflows from the upstream

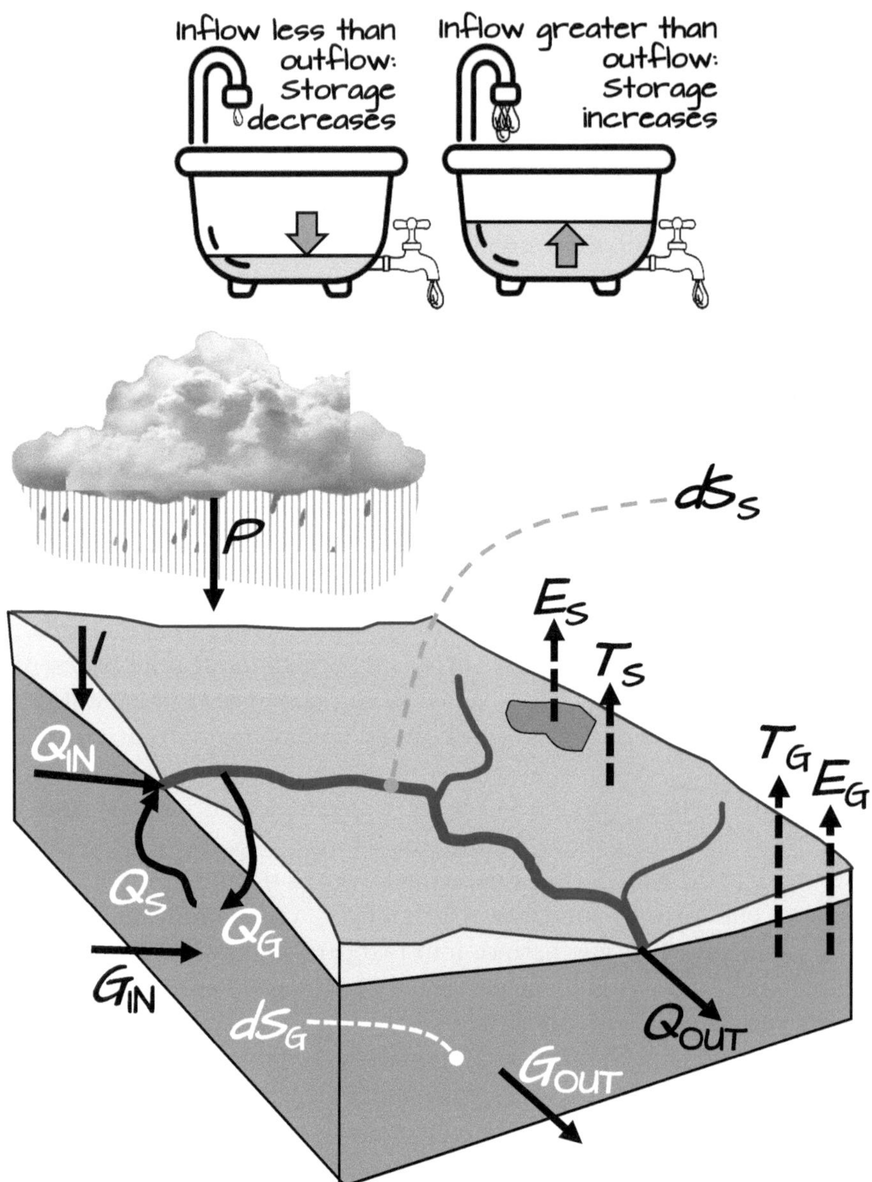

Fig. 4.7 The water balance

aquifers G_{IN}. The negative terms are groundwater outflow to downstream aquifers G_{OUT}, groundwater river supply Q_S, direct evaporation from the soil E_G, and transpiration from plant roots T_G. In practice:

$$dS_G = I + G_{IN} + Q_G - Q_S - G_{OUT} - E_G - T_G.$$

To summarize, the conventional balance equation can be found by merging the two budgets and solely accounting for net fluxes.

$$dS = P - Q - G - ET.$$

The balance between the precipitation input P, the positive income, and the negative losses given by the sum of surface ($Q = Q_{OUT} - Q_{IN}$), groundwater ($G = G_{OUT} - G_{IN}$), and evapotranspiration ($ET = E_S + E_G + T_S + T_G$) outflows results in the storage change of the system $dS = d(S_S + S_G)$.

Compiling the hydrological balance may seem straightforward at first glance because it involves solving an arithmetic equation. Indeed, it can be challenging to measure the relevant fluxes. Moreover, rough approximations are frequently needed in order to address the problem.

Even in what may seem like straightforward scenarios, the complexities of real-world situations are ever-present. Take Khartoum, for instance, where the Blue Nile, originating from the abundant waters of the Water Towers of East Africa, converges with the longer White Nile, whose mysterious headwaters lie in the southern part of the continent (Fig. 4.8).

The Blue Nile contributes roughly 54 billion cubic meters per year (Gcum/y) to the inflows of downstream Lake Nasser in Egypt, where the Nile is impounded by the Aswan Dam. Despite its intermittent flow, the Atbara River, which joins the Nile 300 km downstream of Khartoum, adds about 12 Gcum/y to the inflows at Aswan. In total, Lake Nasser receives approximately 86 Gcum/y of water, primarily from the Blue Nile (63%), with smaller contributions from the White Nile (23%) and the Atbara River (14%). These figures are consistent with the 1959 Nile Agreement which allocated 55.5 Gcum/y to Egypt, 18.5 Gcum/y to Sudan receives, assuming that an estimated 12 Gcum/y is lost due to evaporation and infiltration.

Let us look at Lake Nasser's hydrological system. To assess its water balance, we must monitor various processes such as rainfall across its surface, upstream inflows, and changes in the lake's water level, enabling the calculation of storage volume. The water balance equation helps quantify overall losses due to evaporation and infiltration over different time frames, whether it is a day, a month, or a year. The frequency of measurements determines the temporal scale of analysis.

Drawing from decades of observations, an average annual loss of 12 billion cubic meters (Gcum/y) is noted. However, to fully understand the

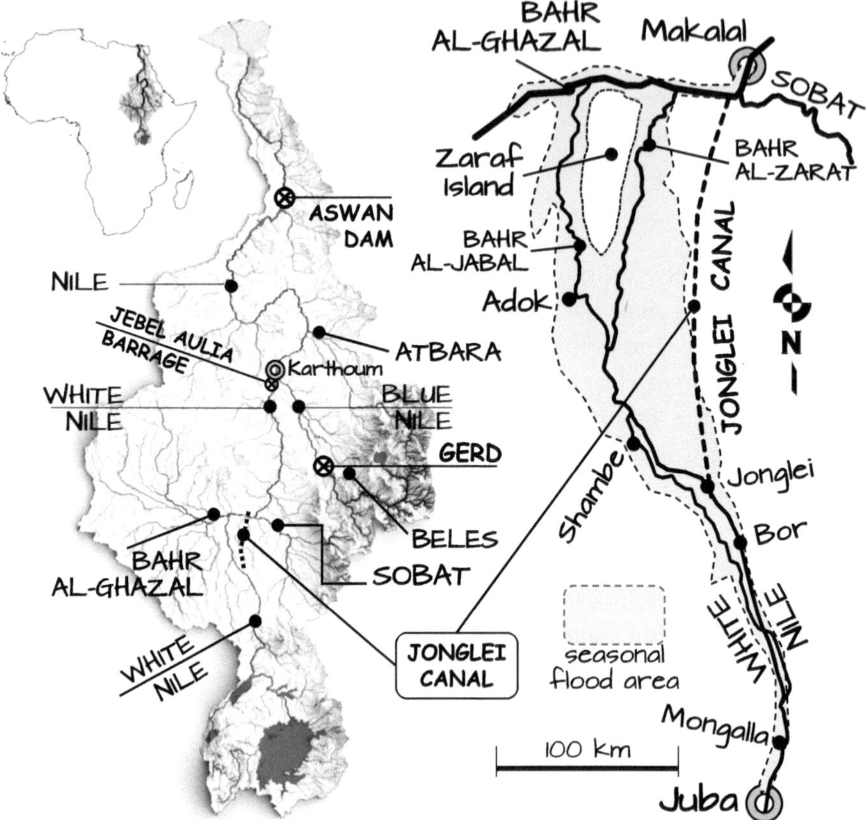

Fig. 4.8 The White Nile in the Sudd

components contributing to this loss—namely, infiltration, evaporation, and seepage—additional data is necessary. Observations of at least two of these processes are required to accurately evaluate the situation. While completing this assessment is feasible, it comes with its fair share of challenges.

The segment of the White Nile stretching from Malakal to Khartoum spans approximately 840 km, with a modest elevation drop of only 13 m across this distance. This reach is characterized by a relatively narrow channel, bordered by vegetation such as papyrus and reeds, along with islands, floodplains, and the banks of the Gebel Aulia barrage.

The tributaries to this segment of the White Nile are small and intermittent, such as the Bahr el Ghazal River, which feeds into Lake No, as well as the Bahr el Zeraf and the Sobat rivers, contributing 4 Gcum/y and 13 Gcum/y, respectively. Despite the contribution by these tributaries, the mean annual flow of the White Nile at Malakal is approximately 30 Gcum/y, whereas it decreases to 25 Gcum/y by the time it reaches Khartoum. This gap can be attributed to losses from evapotranspiration occurring in the floodplains along the White Nile.

Upstream from Makalal, the White Nile virtually disappears into a huge swamp known as the Sudd region in South Sudan. Swamps encompass the riparian portions of the White Nile between Mongalla and Malakal, which spans more than 600 km. This is the largest freshwater wetland in the Nile Basin and among the largest wetlands in the world. The discharge, totaling about 36 billion cubic meters per year (Gcum/y) at Mongalla, reduces by half at Jonglei, situated upstream from the Sabat confluence near Makalal.

In the past, engineers have questioned whether valuable water supplies for that underdeveloped region could be recovered by restoration of the marshes. Sir William Garstin, renowned for designing the first Aswan Dam built at the end of nineteenth century, also proposed the construction of the Jonglei Canal. This canal aimed to connect Mongalla with Makalal, promising benefits such as water conservation, flood reduction, land expansion, and improved navigation links between neighboring countries.

Plans began to take shape much later, in the 1950s, but the Sudanese civil war forced a halt to the project in 1984. Assessing the feasibility of the project needs addressing a crucial question: "Are water losses in the marshes primarily due to evapotranspiration alone, or is infiltration playing a significant role in this loss?"

Evapotranspiration, precipitation, upstream inflows, and downstream outflows of the White Nile must all be estimated in order to solve the water balance equation. Initially, this equation needs to be applied to the various marsh systems, both in a cascading and parallel manner, before assessing the entire system. Potential evapotranspiration significantly exceeds estimated actual losses—approximately 13 Gcum/y from Machar Marshes, 14 from Jebel-Zeraf, and 33 from Bahr El Ghazal (the most relevant). This result suggests that infiltration likely plays a minor role in halving discharge from Mongalla to Jonglei.

The implementation of the Jonglei Canal could potentially increase the annual flow by 27 to 36 Gcum/y, consequently boosting the inflow of Lake Nasser by 7–9%. However, the project is fraught with complex environmental and social concerns, including the collapse of fisheries, depletion of grazing lands, decline in groundwater levels, and potential reduction of rainfall in the region. The dubious benefits of the canal would be split between Egypt and Sudan, with South Sudan likely to bear the brunt of the consequences.

The hydrological balance concept applies to various systems, including artificial lakes, river reaches, swamps, or a combination thereof. Assessing the water balance of the Sudd region results in a favorable outlook for the Jonglei Canal project, aiming to bolster water resources for irrigation and hydroelectric power generation in Sudan and Egypt. This response, however, falls short since it ignores South Sudan's sustainable development as well as more complex environmental and climate-related challenges.

Although the water balance concept can help understanding local or regional hydrology, such as we did for the Sudd, the most appropriate framework is the drainage basin, which refers to a land area where all surface water flows converge to a common point, like a river mouth, or drains into another body of water such as a reservoir, lake, or ocean. The boundary line that delineates one basin from another is known as the drainage divide or watershed (Fig. 4.9). This divide consists of a series of elevated features such as hills and ridges. Various terms are synonymous with drainage basin, including catchment area, catchment basin, drainage area, river basin, water basin, and impluvium.

Numerous options might be taken into consideration if the White Nile water balance is assessed at the basin scale. Countries along the Upper Nile have put forward proposals for water development initiatives, potentially impacting the flow to the Sudd region and the anticipated water savings from the Jonglei Canal. Concerns about environmental impacts have been voiced by local communities, particularly regarding the reduction in swamp size. The Jonglei Canal would decrease the marsh area by 7%, but if upstream development projects are carried out, the area might rise to 16%.

The Origin of Springs

When taking a closer look at the components of the water balance, groundwater emerges as a pivotal player. In Italy, streamflow resulting from direct runoff does not exceed 7% of total precipitation. Rivers, on the other hand, carry 52% of total precipitation, while groundwater contributes 40% of the water supply. This groundwater is replenished by infiltration into the aquifer and subsequently returns this water downstream to the river.

The geological makeup of the Earth's crust, characterized by a combination of permeable and impermeable layers that fold and interact, adds complexity to the aquifers' structure. This means that a given river section may get water from a bigger or smaller basin than what the orography indicates. External inflows and underground losses to neighboring basins may occur depending on the subsoil structure. The effective contributing area and the contributing area arising from the orographic partition can differ in small drainage basins.

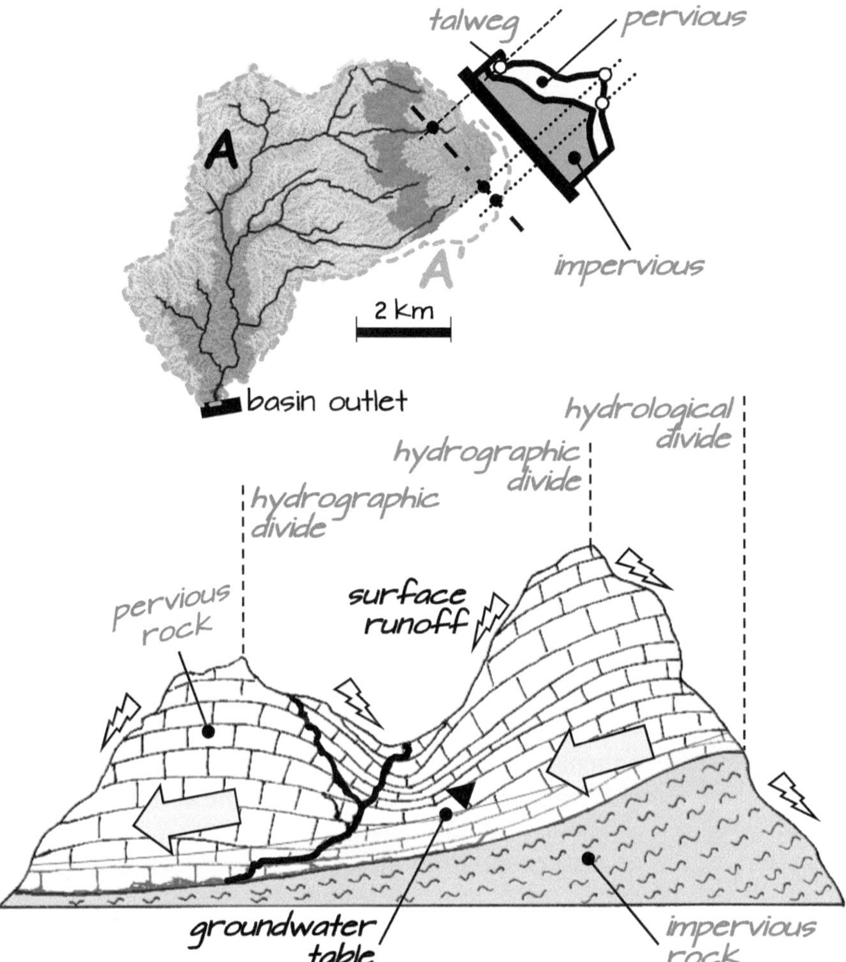

Fig. 4.9 The drainage basin

However, as the basin's size and the river that drains it grows, this discrepancy tends to decrease. The difference becomes negligible with increasing basin size, except for karst hydrology.

Karst systems are formed through the dissolution of soluble carbonate rocks like limestone, dolomite, and gypsum. Flow routes and water storage within karst drainage systems are complex, because greater and lower permeability zones, as well as greater and lower flow velocity zones, are oddly distributed and connected. The water paths are driven by the relative differences in hydraulic head and associated hydraulic gradient between and among these zones.

Water cycle models, which employ the water balance equation across various subsystems and scales, are based on the principle of mass, energy, and momentum conservation, as well as key processes such as evaporation, condensation, and infiltration. In terms of human history concerning water, this knowledge is rather recent. There was no quick or unequivocal consensus over the idea of the hydrological cycle.

To grasp the fundamental principles like the water cycle, scholars first needed to address seemingly simple yet perplexing questions. The crucial inquiry was the origin, location, and timing of springs. Understanding these basic aspects was essential for comprehending the broader dynamics of the water cycle. Can you tell me why, how, where, and when springs come from?

> The origin of water has been the subject of speculation ever since the formulation of tradition and myths regarding the creation of the world. But at the very beginning of Greek philosophy speculation turned from water, as one of the four elements, to the origin of springs.[14]

Throughout ancient times, numerous scholars speculated about the circulation of water. From poets like Homer to philosophers such as Thales, Plato, and Aristotle in ancient Greece, to figures like Pliny, Vitruvius, and Lucretius in ancient Rome. No one could understand the essence of the hydrological cycle, even if some of them managed to create a crude sketch that was not far from actual. It was thought that the ocean provided the springs with their water. In *Hippolytus*, one of Euripides' masterpieces, there is a magnificent rock near Troezen[15] that is said "to distill the water of the Ocean, which pours a fountain very abundant to draw with hydria." However, how is this possible to occur?

At the beginning of the Roman Empire, Seneca devoted the third of his seven books on natural history[16] to the *Forms of water*, which included the circulation of water on and into the Earth. The review scarcely mentions the pluvial theory. Seneca believed that condensed internal vapors fed springs, and rainfall cannot possibly be the source of springs because it penetrates only a few feet into the Earth whereas springs are fed from deep down.

After Seneca there was an almost unbroken silence on the origin of springs until the sixteenth century. The crowd of scholastics of the Middle Ages

[14] Baker, M. N., & Horton, R. E. (1936). Historical development of ideas regarding the origin of springs and ground-water. *Transactions, American Geophysical Union*, Reports and papers, Hydrology: 395–400.

[15] Troezen was an ancient Greek city in eastern Argolis, and the *hydria* was a Greek water-carrying vessel, ceramic or bronze.

[16] Seneca (65 AD) *Natural Questions* (in Latin: Naturales quaestiones).

remained far from a realistic vision of nature. At the time, treatise authors endorsed the ancient theory that springs are formed when underground vapors spring up because of the alembic mechanism. Leonardo da Vinci never wrote a treatise on water, maybe because he had never arrived at a solution, occasionally straying from it entirely and other times coming very close. Despite his keen powers of observation, he repeatedly sketched puzzling diagrams concerning the origins of springs in his notes, confirming the oceanic origin of spring's water (Fig. 4.10).

According to the alembic theory, the springs are fed by the condensation of water vapor that forms in the Earth's core when seawater evaporation seeps through subsurface and beneath crevices. This water would rise to the surface when coming into contact with fire or, to put it more simply, the heat that exists inside the Earth. This mechanism has the name alembic, after the still, which is a tool for artificial distillation. This apparatus has the ability to turn salt water into freshwater and turn grape juice, as well as the juice of other plants, herbs, and berries, into alcoholic beverages like brandy, whisky, vodka, gin, and tequila.

This theory appears utterly absurd today. Even with detailed inspection, it still seems ludicrous. However, it does not minimize—rather, it emphasizes—the importance of the physical processes that were disregarded in the

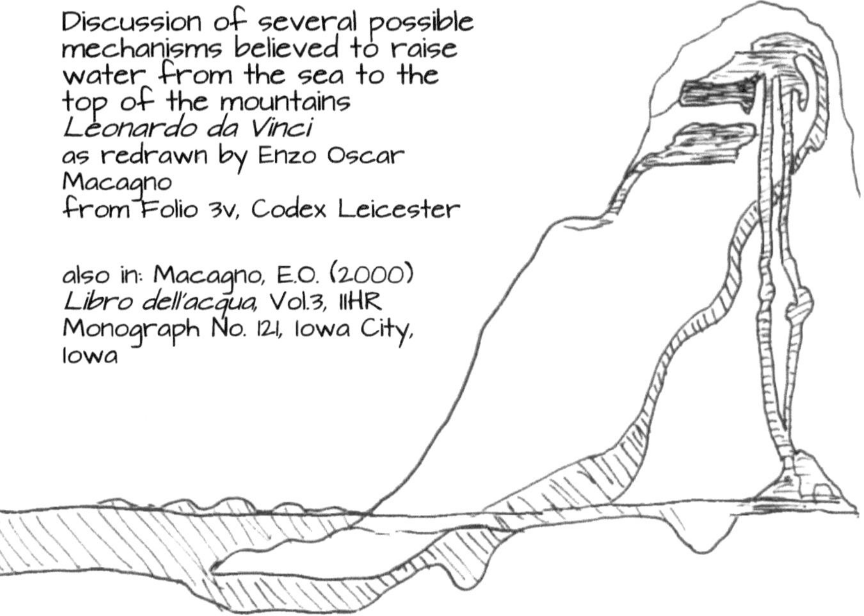

Fig. 4.10 Da Vinci's view of the water cycle

nineteenth and early twentieth centuries, when the hydrological cycle was thought of as a hydraulic machine, a cold machine akin to mill wheels, turbines, and pumps. Everything about thermodynamics was disregarded.

Since solar energy is what propels the constant movement of water from the Earth to the atmosphere through evaporation from water bodies and transpiration from plants, thermodynamics plays a critical role in the transport of water. Heat transfer leads to precipitation, as atmospheric moisture and ice undergo phase changes to form rain and snow, which then descend to the ground through gravity, sometimes triggered by descending winds. The continuous change between the liquid, solid, and gaseous phases that results from thermodynamic processes that absorb a large amount of energy is what drives the hydrological cycle.

Leonardo da Vinci's hydraulic studies of rivers, canals, vortices, eddies, and sea waves paved the way for the principle of continuity of incompressible fluids, as a consequence of mass conservation. He had no idea that the Earth's water cycle could also be explained by using this theory since his drawings frequently reflect different nuances of the alembic theory. However, he remained skeptical of these beliefs and his manuscripts include some grasp of the hydrological cycle.[17] For instance, on Folio 160v of *Codex Atlanticus*, he writes that "the water of the rivers does not originate from the sea but from the clouds."

In the latter half of the sixteenth century, Bernard Palissy—a Huguenot ceramicist, scientist, and hydraulic engineer—presented the earliest compelling argument against the alembic theory, claiming that rainwater fed water bodies.[18] However his findings went largely unnoticed due to the prevailing belief in the endogenic origin of Earth's water. This belief was very popular, especially after when the idea of the central Earth fire core was nearly universally accepted.

At the beginning of the seventeenth century, Thomas Lydiat's cosmogony—a theory capable of proving that the Deluge is a unique, one-time event[19]—was quite popular. One of the most fantastic embellishments was embodied by Kepler, in his *Harmonices Mundi* (1619). The astronomer said "that the Earth forever drinks in water from the sea, like a great beast, digests

[17] Rosso, R. (2020). Leonardo da Vinci's hydraulic heritage: The origin of springs. In: Frega, G., & Macchione, F., a cura di, Tecniche per la Difesa del Suolo e dall'Inquinamento, Vol. 41, 583–592, Cosenza: Editoriale BIOS (in Italian).

[18] Palissy, B. (1580). *The Nature of Waters and Fonteines, both natural and artificial.* Paris: Paris, Chez M. le Ieune, a l'enseigne du serpent, devant le college de Cambray (*De la Nature des Eaux et Fonteines, tant naturelles qu'artificielles*, in French).

[19] Lydiat, T. (1605). *An astronomical lecture on the nature of the heavens and the conditions of the elements.* London: Eliot's Court Press (in Latin).

and assimilates it in its body" adding that "ground-water and springs are the end-products of the Earth's metabolism".[20] Gaetano Fontana used the soul of the geocosmos to explain the emergence of the waters that feed the mountain springs in 1695, one century later than Palissy's treatise.

Although the alembic theory remained popular, Bernardino Ramazzini proposed a more plausible explanation for artesian waters in the Modena area, Northern Italy, suggesting they were sourced from the Apennines range.[21] Nobody paid attention to his conjectures, may possibly be because Bernardino was a physician, the father of occupational medicine, appointed to the first chair of *Theory of Medicine* at the University of Padua. Uncertainty persisted regarding the water's origin, whether it be from rain, snow, or even the sea, throughout the seventeenth century.

Understanding requires measurement, as Leonardo da Vinci argued. What was the required measurement to identify the source of river streamflow? One needed to gauge the quantity of precipitation over a reasonable timeframe and, simultaneously, the total volume of water exiting through the river outlet, alongside fluctuations in water levels in bodies of water, particularly lakes. If the balance sheet returned equilibrium or indicated a deficit, it would indicate that the river's water was solely replenished by precipitation. The precipitation that immediately enters the atmosphere through evaporation from rivers, lakes, and ponds as well as transpiration from plants could have caused the loss. A portion of the rain that infiltrates without resurfacing downstream and is discharged directly into the sea might also be responsible for the loss.

In the late seventeenth century, oceanography presented a challenge to Edmund Halley, the English astronomer who is most known for applying Newtonian principles to determine the periodicity of his named comet. He found that the volume of water discharged from the Mediterranean rivers closely matched the amount of rainfall and snowfall in the regions drained by these same rivers.[22] This finding marked a notable advancement in knowledge.

Around the same period, French naturalist Pierre Perrault—initially a General Tax Collector in Paris and later a scientist—undertook the first serious effort to ascertain whether the amount of rainfall may be adequate to sustain the springs. On his own dime, he financed his research and published the results of his investigations.

---

[20] Baker & Horton. (1936). *Op. Cit.*

[21] Ramazzini, B. (1691). *De Fontium Mutinensium Admiranda Scaturigine Tractatus Physico-Hydrostaticus.* Modena: Typis Haeredum Suliani Impressorum Ducalium (*On the admirable flow of the Modena springs: A physical-hydrostatic treatise,* in Latin).

[22] Dooge J. C. I. (2009). The Concept of the Hydrological Cycle in Britain (1687–1802). *Water International,* 1:4, 18–23.

> It would be necessary to measure or estimate the amount of water in some river as it flows from its source to the point where it joins a creek, and see if rainwater falling around its bed when it is placed in a reservoir, as Aristotle says, would be enough to run the river for a whole year.[23]

Soon after, in 1686, one of the greatest scientists of the day, Abbot Edme Mariotte, published a study on the River Seine and its basin upstream of Paris, thereby validating Perrault's theory on an academic level.[24] When he calculated the upper Seine's yearly flow, he found that it was only about a sixth of the total water falling as precipitation on its catchment area. He astutely deduced that the disparity was attributable to losses incurred through evaporation, transpiration, and irrigation. Thus, he also emphasized the importance of considering the river basin: it was impractical to discuss hydrology without introducing the concept of a drainage basin.

Halley, Perrault, and Mariotte are the pioneering scientists who correctly deduced that precipitation replenished springs, rivers, and lakes. While this notion may have seemed like an advanced concept at the time, it is worth noting that Benedetto Castelli had actually invented the rain gauge nearly half a century earlier, in 1639. Indeed, there had been a lot of alembic water flowing beneath the bridges before the concept of hydrological cycle were disclosed.

These days, it is possible to estimate groundwater rates, streamflow, and the water fluxes between the Earth's crust and atmosphere quite accurately thanks to networks of rain and snow gauges as well as hydrometers in rivers, lakes, and aquifers. Most of the Earth's land is covered by ground monitoring devices. Furthermore, it is possible to measure soil moisture and precipitation using remote sensors on satellites, thereby enhancing the accuracy of these assessments at both basin and global scales. Our capacity to observe has significantly increased because of remote sensing, particularly at the interface between the atmosphere and the oceans. As a result, we now have a very different perspective on what is happening on Earth.

The puzzle of the springs' origin was ultimately resolved by the academic lecture delivered in 1715 by Antonio Vallisneri, a physician and naturalist appointed by the University of Padua.

> The springs are the emergence from the ground of the water penetrated by the rains or following the melting of the snow and, by these hydrometeorological

[23] Perrault, P. (1674). *De l'origine des fontaines*, Paris: Pierre Le Petit (*On the origin of fountains*, in French).

[24] Hubbart, J. A. (2011). Origins of Quantitative Hydrology: Pierre Perrault, Edme Mariotte, and Edmund Halley. *Journal of the American Water Resources Association*, *13*(6), 15–17.

phenomena, a sufficient quantity of water is transferred from the atmosphere to the Earth's surface, enough to justify the amount of the sources themselves.[25]

His presentation drew upon extensive long-term investigations, observations, and conjectures that were rooted in rigorous fieldwork. One crucial piece of evidence supporting Vallisneri's theory was assembled during the summer of 1704. He embarked on extensive travels across Apennine Mountains, spanning from the hills surrounding Reggio Emilia in the Po valley to the native Garfagnana region on the Tyrrhenian Sea.

The new understanding of spring hydrology quickly raised challenges in real life. At the end of the eighteenth century Erasmus Darwin—a physician, natural philosopher, poet, slave-trade abolitionist and grandfather of Charles—made practical use of this knowledge of exposed inclined strata to rejuvenate a bad well by sinking one within it to a stratum that he knew outcropped at a higher level.[26] He also described the origin of springs in his poem the *Botanic Garden* and in his prose work *Phytologia*.

From Data to Predictions

In Padua, Marquis Giovanni Poleni started systematic daily precipitation measurements in 1725. This dataset provides the longest continuous data series worldwide. Italian observatories in large cities gained notoriety in the eighteenth century for their continuous temperature and precipitation monitoring. It wasn't until the early twentieth century that Italy's Hydrological Service established a national framework for precipitation and streamflow monitoring.

In France, the systematic measurement of precipitation started in the late seventeenth century. Louis Morin of Saint-Germaine-en-Laye made one of the first organized attempts to measure rainfall when he started keeping track of daily precipitation observations in 1670. His efforts heralded the beginning of more standardized and sound methods of precipitation measurement in France. A broader nationwide network of meteorological stations was established as a result of the growing sophistication and expansion of these efforts over time. The Paris-Montsouris Observatory provides the longest

[25] Vallisneri, A. (1715). *Lezione accademica sull'orgine delle fontane.* Venice: by Gio. Gabbriello Ertz. Published, with additions, in: A. Vallisneri, *Lezione accademica sull'orgine delle fontane, Seconda Edizione,* Venice: Per Antonio Bortoli, 1726 (*Academic Lecture on the Origin of Fountains*, in Italian).

[26] Baker & Horton. (1936). *Op. Cit.*

precipitation data record in France. Since 1873, this Paris-based observatory has been gathering meteorological data in continuous.

The Reverend John Ray and his associate Francis Willughby kept the first documented systematic records in the United Kingdom in the 1660s. However, the work of Thomas Barker and Richard Towneley in the early 1700s solidified the widespread and methodical recording technique. Since its founding in 1767, the Radcliffe Observatory has been gathering data on precipitation every day starting from 1769.

Johann Kanold, a German scholar, is often recognized as the pioneer of systematic meteorological observations in Germany, which included rainfall and temperature measurements, starting around 1701. His efforts were part of a broader European movement towards regular meteorological observations during that era. By the late eighteenth and early nineteenth centuries, meteorological stations had been established throughout Germany, facilitating more standardized and continuous recording of weather data, including precipitation. The Hohenpeissenberg Meteorological Observatory, located in Bavaria, holds the longest precipitation data record in Germany, having commenced data collection in 1781, thus making it one of the oldest mountain weather stations.

In the United States, the continuous measurement of precipitation began in the late nineteenth century. The longest continuous precipitation record originates from the weather station in New Bedford, Massachusetts, which has been collecting precipitation data since 1849. In 1870, the establishment of the Weather Bureau—now National Weather Service—marked a significant milestone in the systematic observation and recording of weather, including precipitation, across the country.

Monitoring water levels in rivers and lakes is comparatively simpler than measuring rainfall fields. Precipitation is an intermittent process. Predicting precipitation occurrence, rates, and variations in space and time poses significant challenges, often characterized by high uncertainty. In practice, employing a statistical approach is essential for effectively assessing this process. Spatial and temporal scales must be carefully considered since most water processes are best understood using statistical methods.

For instance, the probability of experiencing rainfall events during April stands at 34% in Milan, Lombardy, Italy. But it drops to approximately 25% in August. Moreover, there is a three percent chance that August 15 will be wet in Milan; however, there is a less than 0.05% chance that it will rain between 10 and 11 am. An important consideration for evaluating water fluxes is the time frame of the investigation since the chosen time scale greatly influences the analysis.

Another key aspect is the spatial scale. While there is a nearly certain chance of rainfall in Lombardy during August at the regional level, the situation in Milan, which is less than 200 km^2 in area, presents a stark contrast. Despite its small area relative to Lombardy's 24 thousand square kilometers, the probability of rain in August is considerably higher in Milan than what would be expected based solely on proportional spatial coverage: Milan makes up less than one percent of the whole area!

The superposition principle, which is embedded in the idea of proportionality, is the fundamental concept of linearity. Hydrologic processes are not time-invariant and rarely exhibit linearity. August is not April when it comes to evaluating Milan's rainfall pattern or dressing for a walk in a waterproof jacket. The water cycle is mostly influenced by seasonality, although there are long-term modifications and variations that take decades or even centuries to develop. Long-term variability is seen in long-term precipitation patterns, which can alter river flows, soil moisture levels, and groundwater recharge rates. Examples of this variability include a rise in the frequency of heavy rain events or extended dry spells.

For instance, the precipitation data series of Padua show notable long-term oscillations. However, the examination of long-term data series should carefully evaluate the bias due to changes in funnel position, gauge replacement, low observation regularity, and irregular sampling. Poleni's adjustment of the funnel position in Padua between 1737 and 1742 resulted in somewhat lower amounts being captured. Comparing with contemporary Bologna data revealed that this was caused by an instrumental bias instead of a climate signal.[27]

Long-term changes, increases or decreases, in glacier ice and snowpacks in colder climates can have an impact on downstream river flows and water availability. Rising sea levels may trigger saltwater intrusion into coastal areas, impacting freshwater resources and intensifying the impacts of backwaters during floods. Seasonal water availability can be significantly impacted by changes in snowfall and snowmelt patterns. Local and regional water balances can be impacted by changes in vegetation and land use. Land use modifications can also alter a variety of hydrological processes including transpiration, infiltration, surface runoff, and erosion. Variability in streamflow, both in terms of annual amounts and seasonal distribution, can have profound effects

[27] Camuffo, D., della Valle, A., Becherini, F., & Zanini, V. (2020). Three centuries of daily precipitation in Padua, Italy, 1713–2018: History, relocations, gaps, homogeneity and raw data. *Climatic Change*, *162*, 923–942.

on water supply for human consumption, aquatic ecosystems, and sediment transport.

Because of its key role in shaping ancient Egyptian history, we go back to the Nile River for examining long-term variability of river flows.[28] The oldest of the seven wonders of the ancient world—the great pyramid of Giza that served as the tomb of pharaoh Khufu—was built in 2600's BC about 7 km west of the present-day Nile River. Giza lies on the frontier between the desert and the fertile floodplain east of Giza. It is now accepted that ancient Egyptian engineers exploited a former channel of the Nile to transport building materials and provisions to the Giza plateau. This is supported by palaeoecological analyses that helped to reconstruct an 8000-year fluvial history of the Nile in this area, showing that the former waterscapes and higher river levels around 4500 years ago facilitated the construction of the Giza Pyramid Complex. After a high-stand level concomitant with the African Humid Period, Giza's waterscapes responded to a gradual insolation-driven aridification of East Africa, with the lowest Nile levels recorded at the end of the Dynastic Period. The Khufu branch remained at a high-water level, about 40% of its Holocene maximum, facilitating the transportation of construction materials to the Giza Pyramid Complex (Fig. 4.11).

When examining river streamflow across millennia, long-term hydrological studies show significant changes (Fig. 4.12). The majority of rivers in Europe have demonstrated these noticeable changes over the past few centuries as well, which are also related to the increasing development of water resources systems and associated engineering works. Non-stationary behavior and non-linearity merge together under a complex, eternal braid.

In the twentieth century, Harold Edwin Hurst studied the Nile for the Egyptian government for about 60 years, during which he established the groundwork for an enormous collection of hydrological records and studies. His studies of the size of over-year reservoirs needed to maintain a given yield from Nile flows showed that this was greater than that based on random series predictions[29] (Fig. 4.13). Other natural series supported this discovery, which came to be known as the Hurst phenomenon, and it produced significant advancements in both theory and applied statistics, with a wide range of applications, including financial markets and social phenomena.

Examination of long series of Nile flows, as well as rainfall, temperature, and air pressure, revealed the existence in all these series of periods of generally

[28] Sheisha, H., & 9 co-authors. (2022). Nile waterscapes facilitated the construction of the Giza pyramids during the third millennium BCE. *PNAS*, *119*(37).

[29] Hurst, H. E. (1949). *The capacity needed in reservoirs for long-term storage*. The Nile Basin, Supplement to vol. VII. Government Press, Cairo.

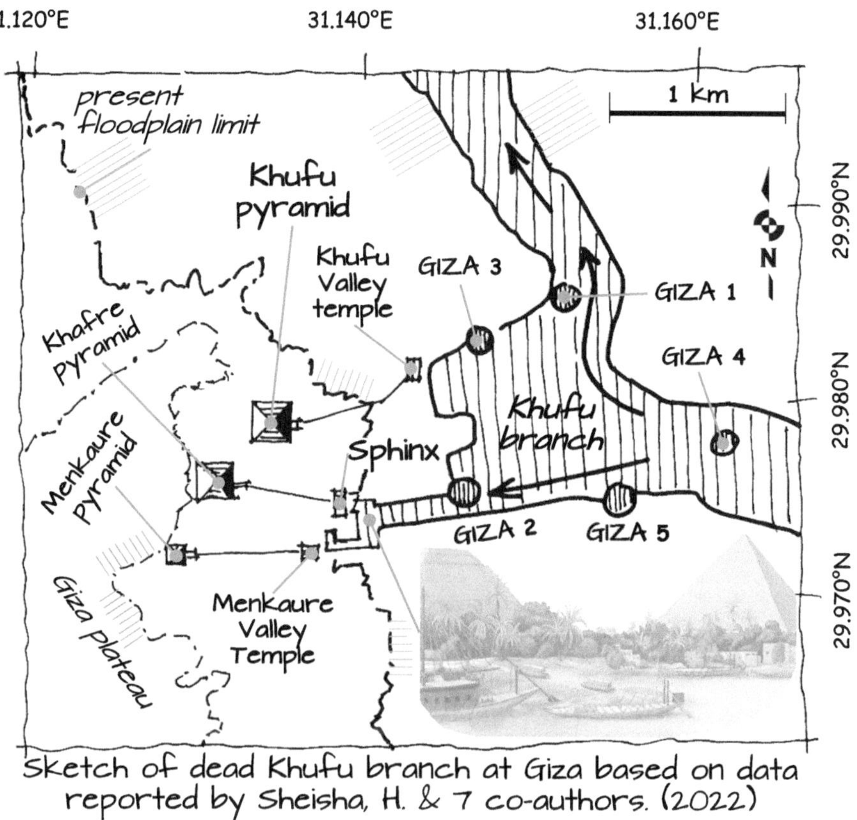

Fig. 4.11 The ancient Khufu branch of the lower Nile

high values and also of low values. In practice, the accumulated departures *R* from the mean did not scale as expected with the length of observations, *N*. For a linear random process, the ratio between *R* and the standard deviation of data, σ, is proportional to the square root of *N*. Hurst found that R/σ is proportional to N^K, but the exponent is quite different from the expected ½, the square root exponent. Hurst's effect, also known as long-term dependence, has significant consequences because it implies that most hydrological time series exhibit persistent trends over time. This means that past values of the series have a long-lasting impact on future values than that expected from a linear approach.

Although time-invariance and linearity are refutable axioms in general, applied hydrology has long relied on these assumptions due to the simplifications they provide. A hydrologic system's output can be predicted using historical input and output data. If a specific input has previously occurred at

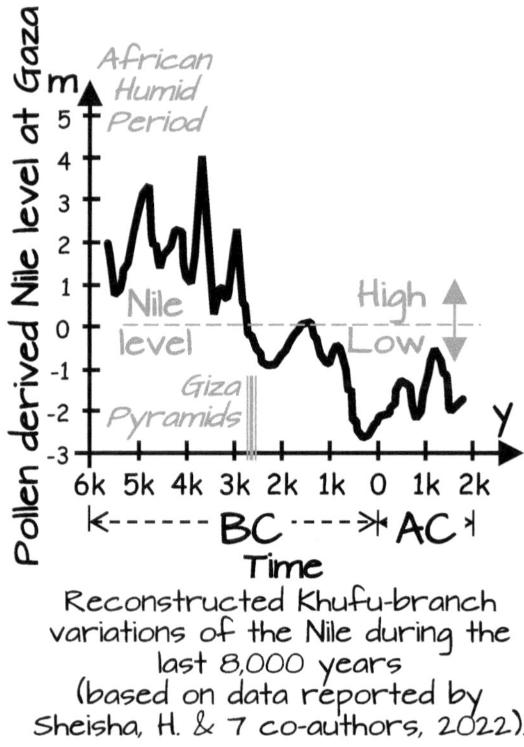

Fig. 4.12 Khufu branch variations over time

some point throughout the record period, then an output for that input can be predicted under the time-invariance assumption. In the absence of the time-invariance assumption, this would seldom be attempted. The further assumption of linearity allows the prediction to be made even though the pattern of input in which we are interested has not occurred in the past.[30]

The linear systems approach usefully describes hydrological processes which exhibit properties such as superposition and proportionality. It is widely used in both deterministic and statistical frameworks. Based on temperature and precipitation data from the past, streamflow can be predicted using statistical linear regression. By fitting a linear relationship between these input variables and streamflow observations, we can forecast future streamflow levels, including their uncertainty range.

Using storm rainfall data as input, the unit hydrograph concept is frequently used for small catchments to describe the time evolution of flood

[30] Dooge, J. (1973). *Linear theory of hydrologic systems, Technical Bulletin 1468*. Washington: United States Department of Agriculture.

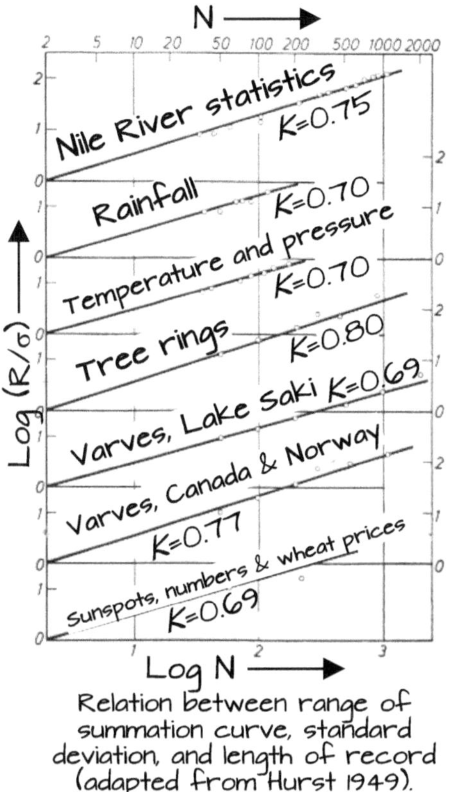

Fig. 4.13 The Hurst's effect

discharge and predict its peak value. The convolution of storm rates with an appropriate linear response function forms the basis of this approach, and this function can be associated with the features of the river network. More frequently, linear transfer function models are employed to simulate the transformation of rainfall into streamflow in watershed systems.

Groundwater flows are predicted using the Darcy formula, the linear relationship between flow velocity and hydraulic pressure gradient. Linear regression analysis can be applied to interpret data from pumping tests conducted in aquifers to estimate parameters such as hydraulic conductivity and transmissivity. Linear advection-dispersion models can be used to simulate the transport of contaminants in groundwater systems. Linear regression or correlation analysis can be employed to analyze the relationship between temperature, snowpack depth, and snowmelt runoff in mountainous regions.

The complexity of real-world systems, which might display non-linear behavior and time-varying dynamics, may not be fully captured by the linear method, despite the fact that it provides a preliminary insight of hydrological processes. The validity of the linear conjecture is limited to appropriate time and spatial scales of study. All things considered, the application of linear system theory in hydrology enables a methodical and quantitative approach to the study of hydrological processes, supporting water management, decision-making, and comprehension of the relationships between various water cycle components.

Where Springs Arise

The ultimate inquiry delves into the origin of a river's source again, merging insights from preceding questions. Vallisneri keeps giving us the initial clue. He was able to record a myriad of phenomena he saw while traversing the Apennines in Central Italy. He intertwined hydrogeological observations with in-depth analyses of the stratigraphic and geomorphological features of the mountains. While hiking up Alpe di San Pellegrino, which is located between Emilia and Tuscany, he saw that springs are not running at the summit of the mountains because there are not any higher areas where water can outflow.

Wandering around the iron mines and some caves in that area, including the chilling Tana Urlante (*howling den*) in Fornovolasco and the Buca di Equi in Fivizzano, he could explore karst aquifers. Fundamental questions arose regarding the interaction between pervious sedimentary layers and impervious sediments, as well as between pervious and impervious rocks. He came to the conclusion that subsurface streams were not sufficiently taken into account by the apparent disruption in the water balance between mountain streams and lower aquifers.

Subsoil's permeability is not uniform nor constant; instead, the strata overlap in intricate ways governed by geology, which includes lithology, stratigraphy, and the structure of deposits and sedimentary formations. When the groundwater table surfaces on the ground, water overflows naturally from the terrain, thus initiating surface runoff. However, the sources differ according to why and how the water table surfaces.

Depression springs are found along a depression, which could be the bottom of a basin, an alluvial valley, or a valley filled with highly permeable soil. Over time, they may vanish due to variations in the aquifer's water level. An example of a depression spring is a small pond that forms in a low-lying area of a forest after heavy rainfall. As rainwater collects in the depression, it gradually seeps into the ground, eventually saturating the underlying soil and forming a spring where water flows to the surface, creating a small stream or pool.

Contact springs form where impermeable rock or soil, known as an aquiclude or aquifuge, underlies groundwater, typically found along hillsides or mountainsides. Fracture springs, also called joint springs, occur when groundwater flowing along an impermeable rock encounters a crack, fault, or joint in the rock. Artesian springs typically manifest in the lowest regions of an area. They arise when the pressure of groundwater surpasses atmospheric pressure, forcing water to surge upward from the ground.

The headwaters are the springs that give rise to rivers, as was previously mentioned. Many rivers have their origins as mountain streams, which start to flow from the icefront of mountains and glaciers downward when ice and snow melt. Prominent examples of icefront origin are the Ganges and Indus from the Himalayan chain, the Rhone at the foot of the Alps, the Columbia River in North America that has its headwaters originating from a glacier in the Canadian Rockies. The Rhine begins at Lake Toma, a small lake in Switzerland near the Oberalp Pass. At 2344 m above sea level, Lake Toma receives water from glacial melting.

The longest river in Italy, the Po, is said to emerge from the subterranean Elisi fields "where the Eridanos stream twists higher through the woods" according to Virgil's *Aeneid*. His virtual student, Dante Alighieri was knowledgeable about the Po's beginnings when he wrote:

> And as the stream, which is the first that eastward
> from Monte Veso takes a separate course
> upon the left slope of the Apennines[31]

Perhaps Dante did not want to contradict his guru, since he still shares Virgil's view about the Po's headwaters. Contemporary Italians locate Po headwaters under a huge boulder at Pian del Re, a large natural amphitheater at 2020 m above sea level, just upstream of Crissolo. In fact, the Po originates at much higher elevation, from the glacier of Monviso.

And what about the Nile's headwaters? In the fifth century BC, the major Greek geographer, Herodotus of Halicarnassus, gave a pioneer location:

> Between Syene and Elephantine, the city of Thebaid, there would be two mountains displaying sharp peaks. The springs of the Nile, which would be at the bottom of the abyss, would flow between these mountains. Half of their water would flow towards Egypt and the wind of the North, the other half to Ethiopia and the south wind.[32]

[31] Dante Alighieri, *Inferno*, Canto XVI, 95–97.

[32] Seppilli, *ibidem*, p. 123.

Catullus was a Latin poet laureate of the late Roma Republic in the first century BC. He lists seven springs. The Nile "colors the meadows" where the outflows occur. This belief captivated the minds of all the poet's contemporaries.

Other powerful men avidly looked for the headwaters of the Nile after Emperor Nero in the first century BC. For instance, Mohammed Ali Pasha, the viceroy of Egypt during the conquest of Sudan, dispatched a naval officer to investigate the headwaters of the Nile. The captain attempted to proceed further upstream in three consecutive attempts (1838, 1839, and 1841). However, he was only able to reach Bedden Rapids in the region of the Fourth Cataract, 200 km downstream Khartoum.

The query had become legendary, and the British Royal Geographical Society supported a number of demanding expeditions in pursuit of these mythical springs. The world was ultimately informed by English explorers Grant and Speke that the Nile originates in Lake Victoria, Nyanza, or 'kinyarwanda' as the locals call it. The year was 1862.

When Gian Lorenzo Bernini created the *Fountain of the Four Rivers*, he knew that the Danube originated from the German Black Forest's River Breg, and the headwaters of the Ganges originate from the Gangotri Glacier in the Indian Himalayas. In fact, the Paraná and Uruguay rivers in South America meet at the confluence of other rivers in Brazil, which is where the Rio de La Plata begins. It is not certain that Gianlorenzo knew the true origin of the Rio de La Plata, but he took it for granted. His only mystery was the location of the Nile's headwaters, which it is now time to reveal.

Indeed, we now know that the Kagera River, which rises in the eastern Congolese mountains close to Lake Kivu, is the actual headwaters of the Father of African Rivers. After that, it enters the large Lake Ukerewe. Explorers named the lake Ukerewe after their queen, Victoria, but today it is known as Nyanza. Only the main island of that huge lake, from which the Nile starts flowing toward the Mediterranean Sea, is identified by Ukerewe.

5

The Governance of Water

What we do to nature or rivers or water,
we do to ourselves.
Donald Worster, The Flow of Empire, *2011*

The Value of Water

In the realm of coveted treasures, few rival the allure of a diamond. Offering this radiant gem to a lady is a splendid strategy to captivate her attention and admiration, setting the stage for a memorable impression. This view does not only apply to ladies, as many gentlemen proudly display a diamond solitaire ring on their little finger. Anyone willing to conquer a companion for life should follow the advice. The diamond is a beautiful stone that is also very useful. It can alter, carve, or work any other material because of its hardness. But life on Earth does not depend on this stone. Humankind would suffer very little without it.

On the contrary, we know that water is essential to life. According to ancient philosophers, water is the pure essence, the primary element of the universe, and the foundation of all things. A brilliant British poet of the XX century, Wystan Hugh Auden, said that millions have lived without love but none without water. Another poet wrote

R. Rosso, *Five Easy Pieces on Water*, https://doi.org/10.1007/978-3-031-69276-5_5

we are water, not the hard diamond
that is lost, not that which rests.[1]

Without water, humanity would not exist. Why is water much cheaper than diamonds, even though the latter is unnecessary and the former is so valuable?

Among the first people to ask the question in the modern age was Adam Smith. He dubbed this seeming contradiction the "diamond-water paradox" and used the "theory of value work" to explain it. The cost of an asset is determined by the labor and resources needed to bring it to market. Since diamonds are significantly harder to find, work with, and market, their price is much higher than that of water.

Costs do not necessarily translate into price; in certain cases, the opposite holds true. A bottle of champagne costs a lot of money. The cost of the land used to grow the grapes, the wages paid to workers, or the long processing all have little effect on the product's pricing. People enjoy drinking fine wine, especially when it is well-advertised to become a status-symbol, which is why it is pricey. The subjective pricing establishes the cost. A diamond is much more expensive than a water bottle due to its symbolic, wholly subjective value.

Not everyone in the world has the means to even buy bottled water. An average of $0.70 is charged for a 1.5-l bottle, according to a September 2022 global survey.[2] A bottle can cost anything from two dollars in the Philippines to less than twenty cents in Tunisia. Prices for bottles vary substantially between nations. In the grocery, a half-liter of bottled water costs 20 cents; in the stadium, it costs 2 euros; and on a low-cost airline, it costs 5 euros. Not to mention some outliers like Kona Nigari, coming from a mountain spring located in deep sea off the coast of the island of Hawaii. It is rumored to have health benefits at the retail price of $420 per 750 milliliters (mL). Much cheaper is Venn from Finland, arguably the purest water on the planet, which sells at €65 per 660 mL. If you're into glamor, Bling H_2O is a good option at a retail price of $38 or more for 720 mL, depending on the bottle's design.

A cubic meter of drinkable water from the home tap costs 20 cents in Milan, although one pays an additional 40 cents for wastewater treatment. A cubic meter tank can hold two thousand half-liter bottles. Paris charges almost 4 dollars per cubic meter. Diamonds cost from 2 to 20 thousand Euro per carat (200 mg) but the retail price can double because of a fine ring's design.

---

[1] Borges, J. L. (1985). Siamo il tempo, in: *Los conjurados*, Buenos Aires: Emecé (*We are the time*, in Spanish).

[2] See: https://www.statista.com/chart/29544/cost-of-a-bottle-of-water-around-the-world/ (Nov 15, 2023)

One argues that the price range of diamond and gold is much narrower than that of water.

For the ancient Romans, water supply was crucial to their society, as Sextus Julius Frontino, the chief of the office of water commissioner in the first century AD, argued with bold comparisons:

> With such an array of indispensable structures carrying so many waters, compare, if you will, the idle Pyramids or the useless, though famous, works of the Greeks![3]

The Roman emperors not only built magnificent aqueducts, but they also used unconventional strategies to guarantee water security, including water trade.

Food trade includes the water required to grow that food. The concept of embedding water within traded commodities and goods is known as virtual water trade.[4] It measures the amount of water utilized in the growth, processing, and delivery of products, as well as their overall production. When a country imports a good, it is also importing the virtual water used to produce that good. The Western Roman Empire's rise and fall serves as proof of the historical significance of water trade across the Mediterranean Sea.[5] A Virtual Water Network of the Roman World was created to face urbanization and climate change. Irrigation advances and virtual water trading through this network increased Roman resilience against inter-annual climate fluctuations. But over time, urbanization triggered by virtual water trade probably made the Empire more vulnerable to climate variability, increased the expense of imports, and brought it closer to the boundary of its exploitable water resources.

Cereals, coffee, beef, leather, and cotton account for the major amount of virtual water that is traded worldwide. In the last forty years, food water trade has more than doubled[6] (Fig. 5.1). By engaging in virtual water trade, certain affluent nations residing in arid regions possess greater food abundance than impoverished nations in wetter areas.

[3] Frontino, S. J. (about 96 AD) *De aquae ductu urbis Romae*: 16 (The Aqueducts of Rome, in Latin).

[4] Allan, J. A. (1993). Fortunately there are substitutes for water, otherwise our hydro-political futures would be impossible. In: *Priorities for water resources allocation and management*, 13–26. London: Overseas Development Administration.

[5] Dermody, B. J. & 8 co-authors. (2014). A virtual water network of the Roman world. *Hydrology and Earth System Sciences*, *18*, 5025–5040.

[6] Tamea, S., Tuninetti, M., Soligno, I., & Laio, F. (2021). Virtual water trade and water footprint of agricultural goods: the 1961–2016 CWASI database. *Earth System Science Data*, *13*, 2025–2051.

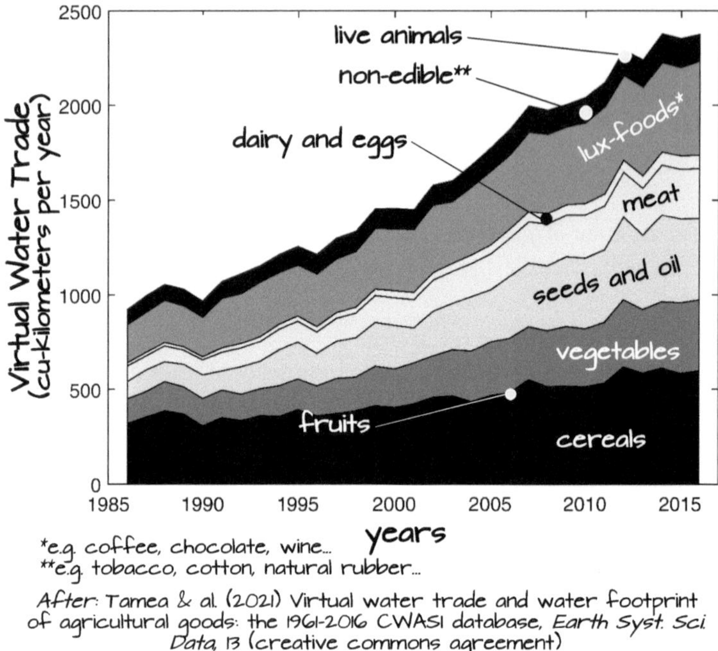

Fig. 5.1 Virtual water trade

Water Demand

These paradoxes illustrate the intricate nature of the relationship between humans and water. The history of mankind's attempts to obtain water for survival is connected to the evolution of water governance. Rainfall has allowed hunter-gatherer men to use wild flora and animals for thousands of years. No water governance was needed. Regrettably, precipitation is quite variable from place to place and from season to season. Generally speaking, this lifestyle soon proved inadequate and unable to sustain populations that were getting denser, more numerous, and more sedentary.

The evolution of food production in favorable habitats caused a profound shift in global social structures around 10,000 years ago. Despite living near the banks of great rivers or having access to rich aquifers, communities often experienced famines. The social mechanisms of those communities underwent a profound renewal to cope with the threat of starvation.

There have been substantial shifts in the connections between food production, arable land, water availability, and societal structure. Human society's dynamics and organization have changed significantly. Even at the local and regional levels, water governance has been progressively and drastically changing.

The changes were centered on the development and application of techniques for irrigation and drainage, water-lifting devices, long-distance water transfer, and storage facilities. The drivers of these changes were the growth of cities, the rising demand for municipal water, and the growth of water-dependent industries and other water-demanding activities. Food production rises when water management is successful, and population growth depends on this. On the other side, failures might have disastrous results.

The rate of population growth and decline was erratic prior to the nineteenth-century industrial revolution. It occurred at slow increase or decrease rates. After this revolution, it began to grow faster and faster. The cities, initially inhabited by a few thousand people, have grown in population. Byzantium had 360,000 residents when became the capital of the Roman Empire in the third century AD. Several hundred cisterns supplied by the Valens aqueduct provided underground water storage (Fig. 5.2). After reaching 500,000 inhabitants, it remained at that level for several centuries. By the time the Roman Empire collapsed in 1453, just 45,000 people called

Fig. 5.2 The Basilica Cistern in Istanbul

Constantinople home. During the Ottoman Empire, the population of the city increased gradually but steadily. It had less than one million residents by 1900, compared to 6.5 million in London. This demographic pattern persisted until 1960, when Istanbul's population began to rise exponentially, reaching over 15 million today. The historic Roman and Ottoman aqueducts continue to supply the city conveying water from the western basins in Europe, but the increased demand has been met by the Melen system, the first transcontinental aqueduct, built to transport Asian water to Europe across the Bosporus Strait (Fig. 5.3). However, recent droughts indicate a further need for increasing municipal water supply.

Once mostly used for livestock and agriculture, the industrial revolution created a new need for water. The counterweight to the extraordinary growth in demand was the reckless assault on the quality of water resources. Industrial and urban pollutants soon threatened not only potability, but also its suitability for industrial and agricultural uses. It took some time to realize that the degradation of freshwater was a direct result of industrial and urban expansion. The more we manufacture by consuming increasingly polluted water, the more challenging it will be to capture clean water supplies. The same

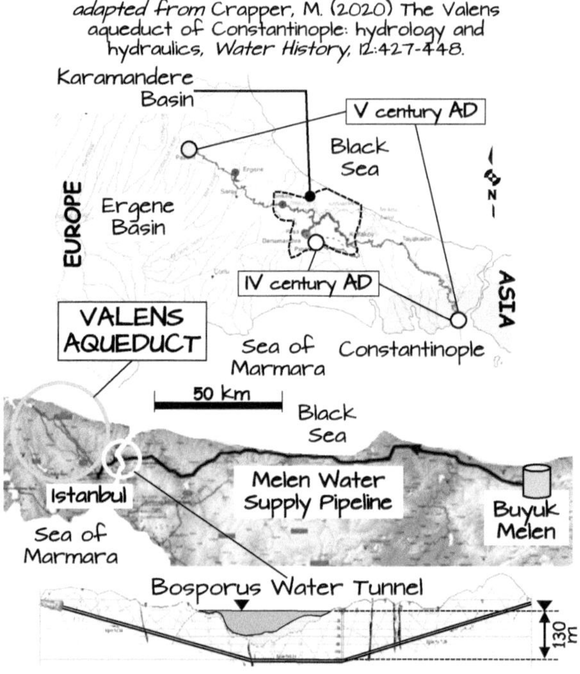

Fig. 5.3 Istanbul's aqueducts

consequence is seen in the uncontrolled growth of water demand by increasing urban populations, which leads to an equally uncontrolled increase in wastewater discharges.

Human kind has been aware of the challenge to water quality quite late, even though it has been clear for centuries. Johann Forster, who traveled with Captain Cook on his second Pacific voyage from 1772 to 1775, observed that the deforestation of Barbados and Cape Verde resulted in severe droughts. In contrast, he acknowledged the English for conserving a large amount 'rain forests' in the highest mountains during their colonization of the East Indies, prohibiting wood cutting to preserve water resources.[7] A half-century later, Alexander von Humboldt, the founder of modern climatology, requested that the Prussian Minister of Commerce preserve the forests of the country. He pointed out that the water level in the Rhine, Elbe, and Oder has decreased systematically over the past forty years due to the myopic deforestation of mountain areas.[8]

It was not until the new millennium that people realized the significant threat posed by water scarcity, even though the United Nations had started to approach the issue of water seriously in 1977 with the declaration of the International Drinking Water Decade.[9] The issue of water scarcity affects all nations, wealthy or not. Demand exceeds availability, and a worldwide water crisis—along with other regional and local crises—should not be completely ruled out. In fact, it might already be underway.

The Water Paradigms

Since water is a finite and a vulnerable resource, yet essential to sustain life, development, and ecosystems, humankind has introduced rules to share and use it throughout history. Water governance has always been approached through the prism of paradigms that reflect prevailing attitudes and

[7] Forster, J. R. (1798). *Beobachtungen und Wahrheiten nebst einigen Lehrsatzen, die einen hohen Grad von Wahrscheinlickleit erhalten haben; als Stoff zur kunfitgen Entwerfung einer Theorie der Erde*. Lipsia: Breitkopf und Hartel.

[8] Humboldt, von A., Letter to the President of Trade Ministry, 17th December 1845; as referred by K. H. Bernhardt in "Alexander von Humboldts auffassung vom klima und sein beitrag zur einrichtung von meteorologischen stationsnetzen". *Zeitschrift fur Meteorologie, 34*, 213–217, 1984.

[9] See: United Nations Water Conference, Mar del Plata, Argentina 14–25 March 1977; and International Drinking Water Decade, 1981–90.

practices.[10] These paradigms canonized and codified current mental structures through social communication, behavior, and interpretation processes.

The oldest is the *spiritual-religious paradigm*, which is deeply ingrained in mythology, tales, and religious doctrines all around the world. The *paradigm of hydraulic engineering* dates back to some centuries BCE, and the development of Islamic water technology significantly reinforced it much later. With the onset of the industrial period, this worldview gave way to the *scientific paradigm* that views water primarily as a chemical and physical material whose properties are investigated by science.

By the end of the twentieth century, the *economic–financial paradigm* became the prevailing archetype, triggered by the rise of the industrial and hydraulic engineering paradigm. The *ecological paradigm* is a somewhat counter-paradigm of the latter because it emphasizes sustainability as it relates to ecology, environmental ethics, health, and spirituality. It is aligned with the spiritual-religious paradigm in this way.

An *aesthetic-recreational paradigm* was created by royals, state rulers, and nobility using gardens, water fountains, baths, and spas to enjoy and flaunt their rank. This paradigm has guided royal families throughout the Renaissance and Modern Eras, beginning with the Roman Emperors and Ottoman Sultans.

The *scientific-health paradigm* treats water as a chemical given its potential to harbor pathogens. Established in the seventeenth century, it comes from the sanitary paradigm that Vitruvius raised in his masterpiece *De Architectura* twenty centuries before, let alone the Paracelsus heritage of early sixteenth century alchemy.

Although the *legal and ethical paradigm* has existed since the establishment of national states, it has mostly taken over in the last several decades as disputes between users inside and between countries have begun to intensify.

Guidelines for Water Management

New principles of water management have been designed in the last 30 years to ensure the equitable and efficient management and sustainable use of water. The Dublin Principles on Water and Sustainable Development were the first global attempt to state the main issues and purpose of water governance.

[10] Hassan, F. (2011). Water: History for our times, IHP Essays on Water History, Vol. 2, Paris: UNESCO Publishing.

It is possible that good intentions do not always provide the desired results. In practice, water privatization was promoted by the 1992 Dublin Statement, which said that water should be treated as an economic good. This had a detrimental effect on human rights, particularly in poor nations. After this declaration, the hydraulic engineering paradigm of water governance, which peaked during the twentieth-century era of building large dams worldwide, was first supplemented and subsequently superseded by the financial-economic paradigm in the final decade of the twentieth century.[11]

The OECD released its "Principles on Water Governance" in 2015, many years after the Rio de Janeiro Earth Summit, the 1992 United Nations Dublin Conference, and the Dublin Declaration.[12] More than 170 governments and stakeholder groups, including seven non-OECD members, 140 stakeholder groups, and members of the Organization for Economic Cooperation and Development, supported these statements. A *stakeholder*—namely, someone carrying a stake, for example the wooden one needed to kill a vampire—is a person or an organization that has a legitimate interest in a project or entity. It has gained wide acceptance in business practice, strategic management, and corporate governance.

> A stakeholder can be an individual or a group, with the word 'anyone' inviting us to draw our net as widely as possible. And any interest means that they can be interested in what you are doing, how you are doing it or its outcome.[13]

Since water is an economic good, one must adopt the economy vocabulary when discussing water. The OECD guidelines offer 12 *must-do's*, that are requirements that governments should meet to design and implement effective, efficient, and inclusive water policies. These principles are the following.

1. Clearly allocate and distinguish roles and responsibilities for water policymaking, policy implementation, operational management, and regulation, and foster coordination across these responsible authorities.
2. Manage water at the appropriate scale(s) within integrated basin governance systems to reflect local conditions and foster coordination between the different scales.
3. Encourage policy coherence through effective cross-sectoral coordination.

[11] Petrella, R. (2001). *The water manifesto: Arguments for a world water contract.* London: Zed Books.

[12] https://www.oecd.org/governance/oecd-principles-on-water-governance.htm

[13] Clayton, M. (2014). *The influence agenda. A systematic approach to aligning stakeholders in times of change.* London: Palgrave Macmillan.

4. Adapt the level of capacity of responsible authorities to the complexity of water challenges to be met.
5. Produce, update, and share timely, consistent, comparable, and policy-relevant water data.
6. Ensure that governance arrangements help mobilize water finance and allocate financial resources efficiently.
7. Ensure that sound water management regulatory frameworks are effectively implemented in pursuit of the public interest.
8. Promote the adoption and implementation of innovative water governance practices across responsible authorities.
9. Mainstream integrity and transparency practices across water policies for greater accountability in decision-making.
10. Promote stakeholder engagement for informed contributions to water policy design and implementation.
11. Encourage water governance frameworks that help manage trade-offs across water users.
12. Promote regular monitoring and evaluation of water policy where appropriate.

While these guidelines are appropriate, the overall must-do's approach prioritizes managerial and budgetary concerns. What about the fundamental issue raised in the same year by the Holy Father Francis with his *Encyclical Letter Laudato Sì* when stating that climate is a "common good belonging to all and meant for all"? Water is a common good, not a commodity or merchandise. And not a good only subject to market laws.

> It is necessary to develop financing plans as well as wide-ranging water projects. This resolve will lead to overcoming the notion of turning water into a mere commodity, regulated exclusively by market laws.[14]

Management and governance are often associated with water randomly. Despite their close relationship, water management and water governance have different meanings.[15] The first refers to the tasks associated with managing, overseeing, and regulating water resources. Water resource allocation, planning, and optimization fall under this category. On the other side, water governance is concerned with the enabling environment in which water management actions take place. It covers the broad plans, policies, strategies,

[14] Message of His Holiness Pope Francis on the occasion of *World Water Day 2019*.

[15] Ashish Pandey, Mishra, S. K., Kansal, M. L., Singh, R. D., Singh, V. P. (2021). *Water management and water governance: Hydrological modeling*. Heidelberg: Springer.

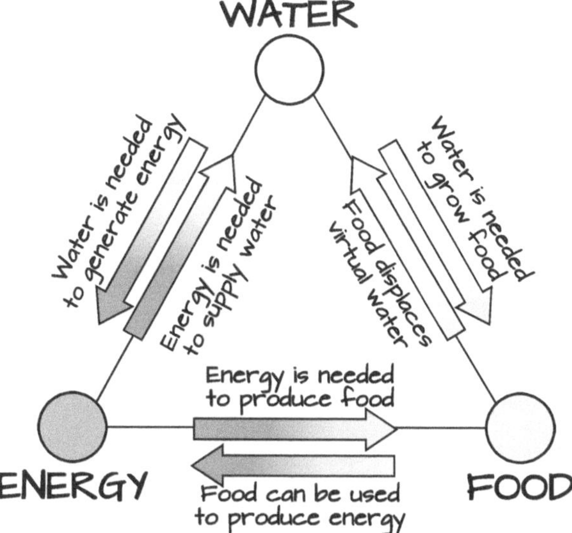

Fig. 5.4 The water-food-energy nexus

budgets, and incentive systems that are related to or have an impact on water resources. In addition, planning, decision-making, monitoring, and institutional structures and laws are involved. Put differently, it concerns who gets which water, when, and how much of it, as well as who is entitled to water and associated services and benefits.

The FAO's Water Governance initiative is a far more comprehensive plan than that by OECD. The Food and Agriculture Organization (FAO) is focused on several aspects of water governance, such as managing water quality, groundwater governance, irrigation governance, water tenure, governance in river basins and watersheds, and irrigation governance. Food security is prioritized by FAO in the global water discussion.

Both the OCSE and FAO approaches are mostly influenced by the *economic–financial paradigm* that watches over ethical and ecological issues. Moreover, energy cannot be neglected (Fig. 5.4).

> Water, food and energy form a nexus at the heart of sustainable development. Agriculture is the largest consumer of the world's freshwater resources, and water is used to produce most forms of energy. Demand for all three is increasing rapidly. To withstand current and future pressures, governments must ensure integrated and sustainable management of water, food and energy to balance the needs of people, nature and the economy.[16]

[16] https://www.unwater.org/water-facts/water-food-and-energy

To address this challenge, management practices should be guided by governance principles that are shared globally.

Some Lessons from the Nile

The Nile River is the 'father of African rivers' that has shaped the landscape and livelihoods of communities along its banks. Originating from rainfall in Equatorial Africa and Ethiopia, the Nile's seasonal floods created a narrow yet fertile floodplain, establishing a rhythm of life for Egyptian farmers. The streamflow spread naturally through the floodplain changing constantly as the Nile changed its course (Fig. 5.5).

Agricultural villages and riverine areas faced threats from both high and low flows. To mitigate flood risk, people built embankments and drainage systems to manage excess water. They also extended canals to supply water to peripheral areas and cope with droughts. Over time, these facilities were enhanced with the building of dams to enable the retention of floodwater, benefiting downstream farmers. The key was collaboration as communities worked together to excavate canals, build dams, and fortify embankments, thereby adapting to and managing the dynamic nature of the river.

Nile water was valuable for purposes other than agriculture and municipal supply. Fluvial navigation was crucial in building the pyramids, as it enabled the transfer of supplies and construction materials (Fig. 4.10). The pyramids were typically positioned west of the Nile because the divine pharaoh's soul was expected to join with the sun during its descent.

Despite agriculture being introduced to Egypt 7000 years ago, a state-level organization was not established until twenty centuries later. This state was ruled by a king believed to possess divine lineage, responsible for tax collection and overseeing royal funerary and temple projects. Until roughly 1880 BCE during the Middle Kingdom, there is no hard proof that the state was involved in creating or managing significant irrigation schemes.

The first large-scale project, a dam known as Sadd el-Kafara, was built across a desert wadi[17] known as Wadi Garawi about 4500 years ago. However, prior to its completion, it was destroyed by a major flood. The purpose behind its construction remains unclear, though it is speculated to have been a failed attempt to build a dam using pyramid-building techniques.

During the Middle Kingdom, 4200 years ago, the rulers faced a major challenge due to severe droughts. Water from a branch of the Nile River was

---

[17] "wadi" id the Arabic term referring to both a river valley and an ephemeral stream.

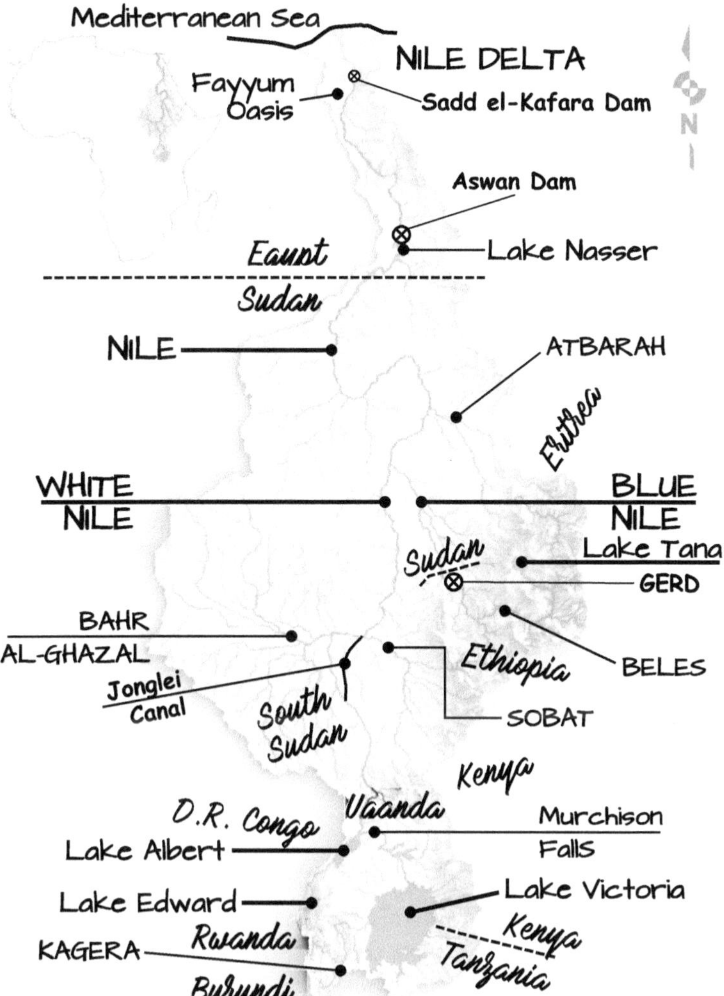

Fig. 5.5 The Nile River

redirected into a depression in the Faiyum region as part of the Middle Kingdom project, which was built between 1880 and 1800 BCE to face the drought threat. This created a prosperous agricultural pharaoh's estate near Memphis, the capital. A few years later, a flood ruined the dam, hastening the collapse of the centralized reign as a direct consequence of prolonged droughts. This dam was restored by the Ptolemys around 300 BCE, and it has undergone multiple rehabilitations over time. It is still a component of the Egypt's complex water management system including dams, barrages, canals, and drains.

The development of water governance along the Nile River is intertwined with the region's historical narrative. Throughout the colonial period, control over water resources in the Nile Basin was exercised by colonial powers. Negotiations and agreements regarding Nile waters were facilitated with the assistance of Great Britain, yet were not inclusive of most riparian states.[18] These agreements largely favored Egypt's rights to the Nile's waters.

The initial treaty, forged in 1929 between Great Britain and Egypt, established frameworks that heavily favored Egypt's interests. During this time, Britain also represented Uganda, Kenya, Tanzania, and Sudan. In 1959, another significant agreement was reached between Egypt and Sudan, which allocated the lion's share of the Nile's waters to these two nations. However, the broader community of riparian states did not ratify this agreement, rejecting the 1929 treaty altogether in 2010.

Following the colonial period, new challenges emerged in Nile water governance. Upon gaining independence, riparian states began asserting their rights over the river's resources, sparking tensions and disputes. In response to the challenges, the Nile Basin Initiative (NBI) was launched in 1999 by the nine countries sharing the river: Egypt, Sudan, Ethiopia, Uganda, Kenya, Tanzania, Burundi, Rwanda, and the Democratic Republic of Congo, with Eritrea as an observer.

The primary objective of the NBI is to foster cooperative development, ensuring equitable sharing of socioeconomic benefits and promoting regional stability. A Cooperative Framework Agreement was signed in 2010 to guarantee each country the ability of developing water projects without the prior consent of Egypt. However, Egypt has chosen to refuse this agreement, leading to ongoing disputes within the region regarding water governance.

Water management strategies in the Nile Basin exhibit a wide array of approaches, reflecting the diverse geographical and climatic factors at play. Irrigation stands out as the key practice, notably in Egypt and Sudan. Traditional irrigation methods, honed over centuries, coexist with modern techniques aimed at enhancing water efficiency. In Ethiopia, a widespread practice is community-based rainwater harvesting, wherein communities gather and store rainwater for later use, particularly during periods of drought.[19] Small-scale irrigation systems are operated across various regions within the Nile Basin, often managed at the community level. These systems

[18] Tayia, A., Barrado, A. R., & Guinea, F. A. (2021). The evolution of the Nile regulatory regime: A history of cooperation and conflict. *Water History, 13*, 293–317.

[19] Kloos, H., & Legesse, W. (2010). *Water resources management in Ethiopia: Implications for the Nile Basin*. Amherst: Cambria Press.

play a vital role in sustaining local agriculture and ensuring food security within the region.

Reforestation is an imperative practice in the medium and upper Nile Basin. By planting trees, communities can help to save water and maintain the health of local ecosystems. Soil and water conservation practices prevent soil erosion and preserve the quality of water resources. These practices can include contour plowing, terracing, and the construction of small reservoirs.[20]

Effective flood control measures are essential in regions that are at risk of flooding. Since ancient times, Nubia and Egypt have experienced beneficial periodic flooding, which is celebrated by the Egyptians with a 2-week long annual holiday called Wafaa El-Nil that starts on August 15th. However, the eastern Nile region—which includes Egypt, Ethiopia, South Sudan, and Sudan— is vulnerable to both floods and droughts, which can have a negative impact on people's lives and property.

Uncontrolled flood events diminish agricultural productivity, reduce incomes, heighten the risk of sickness, disrupt education, and damage public and private assets. At the end of summer each year, monsoon rains pour into the Ethiopian Highlands and flow down to the Blue Nile and White Nile impacting hundreds of villages along the banks, particularly during wet years. Coping with severe flooding, such as the 2021 catastrophe in Al Jabalain district of White Nile State in Sudan, requires effective flood control measures. In Ethiopia, sporadic flash flooding impacts several riparian areas of Lake Tana, while the lowlands of the Baro-Akobo sub-basin are partially submerged by floodwaters annually.

Building levees, floodwalls, and reservoirs are examples of structural flood control mitigation techniques. Non-structural strategies include better risk mapping, early warning systems, planned flood routing in specific areas, and increased disaster preparedness and response. All these measures need an approach at the basin scale that also prevents from transferring flood hazard downstream.

Hydropower is a major issue in Nile water governance. A milestone in was the 1959 Agreement, the groundwork for sharing the benefits from the construction of the High Aswan Dam, initiated in 1960. Lake Nasser reached its maximum reservoir capacity of approximately 169 billion cubic meters by 1976. The massive rock and clay dam has a twin goal, downstream irrigation,

---

[20] Allan, J. A. (1994). The Nile Basin: Water management strategies. In: *The Nile: Sharing a scarce resource – A historical and technical review of water management and of economical and legal issues*, edited by P. P. Howell and J. A. Allan, Cambridge: Cambridge University Press.

and energy generation, with an annual output of 10 TWh, more or less the annual electricity use of four million of Italian houses.

The site for the High Aswan Dam was chosen near the existing Low Aswan Dam, which was completed in 1902. It was not the earliest project to impound Nile's streamflow at Aswan. An attempt to build a dam near Aswan dates back to the eleventh century when the Fatimid Caliph commissioned the Arab engineer Alhazen to explore how to control Nile floods. This included the potential construction of a dam in Aswan. But Alhazen's fieldwork showed that such an idea was unfeasible.

If the Aswan dam agreement was not too unfair for Sudan and did not affect the upstream countries, the cross-border implications of the Grand Ethiopian Renaissance Dam (GERD) are much more severe. Given Ethiopia's position as the 'Water Tower of Africa', the GERD is a massive project intended to produce more than fifteen Terawatthour per year of electricity as part of the Ethiopian government's ambition to boost national hydropower capacity. However, his idea is not supported by the downstream neighbors, including those who share the Omo-Bottego River, the main tributary to Lake Turkana in Kenya.

The construction of the GERD, a gravity dam on the Blue Nile River in Ethiopia, started in 2011. Located 14 km east of the Sudanese border and 40 km downstream of the Beles River junction, its primary objective is energy generation, boasting an installed capacity exceeding 5 GW. With a yearly production of 16 TWh, the dam is now Africa's largest hydroelectric power plant and among the top 20 worldwide. The dam flooded 1680 km^2 of forest in northwest Ethiopia after its filling finished in 2023, displacing some 20,000 people from their homes, and creating a reservoir of about 74 billion cubic meters of water, a volume approaching the entire annual flow of the Blue Nile at the Sudanese border.

What about downstream? Egypt seeks to ensure water security for its 110 million residents, as the Nile supplies 90% of the country's freshwater. Sudan, as the main and immediate stakeholder, is fighting for its survival and prospects for advancement. Ethiopia, on the other hand, has to supply more than 120 million people and drive its own development with low-cost energy, and perhaps become the continent's biggest supplier of renewable energy.

The planning process for the GERD has been top-down and unilateral. The public and communities affected by the dam have not been adequately engaged to provide meaningful input or critique the project. Changes to the project are met with resistance from the Ethiopian government. Damming a shared river unilaterally does not meet the best practices for transboundary water management. The hydrological analysis has been very basic thus far, lacking the level of detail, sophistication, and reliability expected for a project of this magnitude and regional impact. Therefore, there are major concerns regarding the unexplored effects of the GERD on the flow of the Nile River. Criticisms of the GERD center around its potential to jeopardize downstream water security, livelihoods, and ecosystems due to decisions made without adequate consultation or consideration of downstream impacts.

The outdated governance treaties cannot accommodate GERD management policies. However, the new dam could be viewed as a chance to usher in a fresh era of collaboration within the Nile Basin. It is both an opportunity for economic development and a source of dispute. The dam has the potential to bring broader benefits to the region by providing electricity to Ethiopians and generating excess energy available for export to neighboring nations, since the GERD can increase total electric energy production of Africa by 2 percent.

The GERD emphasizes the necessity of riparian states' collaboration to guarantee a fair distribution of water resources. As Africa's largest hydroelectric project, it shows how renewable energy generation can help reduce the continent's energy poverty. But it raises important issues about environmental sustainability and social justice as well. Environmental impact assessments and inclusive decision-making processes are needed to approach the displacement and relocation of local communities and the long-term hydrologic, ecologic, and social effects.

Another project with transboundary effects is the Sudan Jonglei Canal. This massive and challenging engineering project, which is intended to divert water from the White Nile, carries far-reaching consequences. Although fully located in South Sudan, it directly impacts regional politics. Reducing water loss in the Sudd, one of the planet's largest wetlands situated in South Sudan, is the main goal. The dense swamp vegetation of the Sudd, coupled with the evaporation process, results in a significant loss of the Nile's waters.

The canal, spanning approximately 360 km, aimed to serve as a shortcut for the Nile, circumventing the Sudd wetlands (Fig. 4.8). This would boost the quantity of water accessible downstream for Sudan and Egypt. Although its construction commenced in the 1970s, it was halted due to the outbreak of civil war in Sudan. As of April 2024, the project is still being discussed, but it has never been resumed.

The Jonglei Canal project poses a number of intricate governance issues. Every nation along the Nile depends on its waters. Any alteration to its flow has the potential to escalate tensions regarding water rights, particularly among nations like Egypt, Sudan, and South Sudan. There are concerns about fair water sharing because the canal may help Egypt and Sudan at the expense of South Sudan.

Changes to the Sudd's natural water flow would have a big environmental impact. The Sudd serves as a biodiverse ecosystem and functions as a natural water purifier. Modifications can jeopardize local wildlife and the livelihoods of communities that depend on it. The canal might upend the way of life for the local communities in the Sudd, as their way of life is strongly linked to the wetland. Concerns regarding livelihood loss and displacement raise questions about compensation and consent.

With South Sudan's history of civil war, large-scale projects such as the Jonglei Canal bear significant implications for regional stability. Since it was started prior to South Sudan's independence, the project intertwines with issues of national sovereignty, as recently stated.

> The Nile waters and the Sudd wetlands constitute an important resource to the existence and survival of the people of South Sudan. For this reason, tampering with the Sudd area therefore constitutes a national security threat for the people of South Sudan.[21]

Moreover, the engagement of international actors, including financial support and technical assistance, leads to additional complexities. Such involvement brings with it geopolitical interests and pressures. Effective project management demands robust legal and policy frameworks addressing transboundary water governance, environmental preservation, and the rights of impacted communities. These frameworks need negotiation between all stakeholders as well as diplomatic and legal acumen to navigate the complexities involved.

Principles of Water Governance

The never-ending water story of the Nile River claims for a deeper insight of principles of water governance based on shared axioms. This collection of axioms is aimed at directing the choices and actions of different stakeholders involved in water-related issues. It should reflect the economic, social, environmental, and ethical dimensions of water, aiming at ensuring its sustainable and equitable use. The core of water governance is based on the human rights to water, the ecosystem focus, the principles of shared integration, precautionary, responsibility, and subsidiarity, along with the common good concept. Any water management guideline must be implemented under these fundamental principles.

[21] National Salvation Front. (2022). Jonglei Canal. NAS/&OSM/11, tenth May.

The Principle of Human Right to Water

The concept of the human right to water relies on the recognition that water is essential for the realization of other human rights, such as the right to life, health, food, education, and dignity. It also implies that everyone should have access to an adequate supply of safe, physically accessible, fairly priced water for residential and personal use.

States are required by the human right to water to uphold, defend, and grant this right to their people. States have to abstain from obstructing or restricting people's access to water. It is their duty to prevent and remedy any harm brought about by third parties or natural factors that affect the amount or quality of water. In order to guarantee that everyone can actually enjoy this right, they must take proactive steps crating institutional, governmental, and legal frameworks to impellent and monitor the exercise of this right.

The human right to water includes the empowerment and participation of individuals and groups in water-related decision-making. For individuals to be capable of voicing their views on water policies and services, they must have access to information and education on water-related topics. It is imperative that whoever is in charge of supplying or controlling water be held accountable. Effective remedies and justice should be available to people in the event of water violations or disputes.

States are obligated to respect, protect, and fulfill this right for their citizens, as recognized by the United Nations General Assembly in 2010.[22] The right to water is guaranteed by the constitution in about 40 states, including South Africa, Kenya, Uganda, and Zambia in Africa and Brazil, Uruguay, Mexico, Colombia, and Peru in South America. The Constitution of the United States only obliquely recognizes the right to water as a subset of the right to a healthy environment. The right to water is not mentioned in the Chart of nearly all OECD nations.

The Ecosystem Principle

Water is a basic component of the natural world and sustains a number of ecosystems that offer essential services for human survival and well-being. An ecosystem-based strategy considers the interactions between water and other elements of the environment, including air, land, biodiversity, and climate. This principle states that water management must ensure the ecological

[22] Human Rights to Water and Sanitation, UN-Water A/RES/64/292: see https://www.unwater.org/water-facts/human-rights-water-and-sanitation

integrity and resilience of water resources and ecosystems are maintained or restored.

The ecosystem approach can improve the productivity and sustainability of agricultural systems by enhancing the natural mechanisms that promote crop growth, such as soil fertility, water retention, pollination, pest control, and nutrient cycling. Improved trade-offs and synergies are achieved by integrating ecosystems, water resources, and energy production. Urban areas can become more resilient and livable by introducing natural features into built environment.

Maintaining or restoring the ecological integrity of water supplies and ecosystems has both advantages and drawbacks. The benefits include enhancing the delivery of ecosystem services, reducing disaster risk, and promoting the conservation of biodiversity. The trade-offs include social and economic costs, conflicts between different goals, and the challenge of facing constraints and limitations. Given the uncertainty embedded in the complexity of natural systems and the lack of adequate knowledge, data, or ability, significant economic investments may be required.

The Shared Integration Principle

The sharing concept deals with Integrated Water Resources Management (IWRM). This approach promotes the coordinated use of land, water, and related resources to enhance social and economic well-being without jeopardizing vital ecosystems. IWRM needs a comprehensive perspective that accounts for the relationships between water and other environmental, social, ethical, and technological components to design, implement, and evaluate water policies and actions. Accordingly, it can promote sustainable development by balancing all implications of water use.

This approach can be applied in a variety of contexts and situations. Within a transboundary area like the Nile River, IWRM must encourage nations to work together to share water resources. This is not a utopia. Fourteen Pacific Island countries have adopted IWRM at the national and regional levels to tackle water governance, encompassing matters pertaining to water quantity and quality, sanitation, and climate change adaptation.

IWRM can help managers and citizens to improve preparedness to cope with droughts and floods. Flood vulnerability can be addressed by integrating structural and non-structural measures. Integrating different sources can enhance municipal water security as pioneered by the Four National Taps of Singapore, a comprehensive strategy to water management. This has improved the livability and resilience of the city through the diversification and optimization of water supply by integrating local catchment water, imported water, high-grade reclaimed water, and desalinated water.

The inclusion of Nature-based Solutions (NBS) in IWRM is another pillar. These activities address a range of environmental, social, and economic issues in urban environments by using or mimicking natural processes.[23]

The Precautionary Principle

Natural or human-induced factors, such as pollution, overexploitation, or climate change can lead to various risks and threats to water resources and ecosystems. As a result, uncertainty and vulnerability rise. Water governance should ensure proactive steps to foresee, stop, or lessen any possible harm to ecosystems and water resources before it occurs.

When developing precautionary water strategies, one should include scientific uncertainty and account for the lack of conclusive evidence about the impacts of substances or special activities on the environment and human health. This need can be met by implementing proactive and preventive steps to safeguard ecosystems and water resources from a possible harm, instead of using a reactive and corrective strategy that waits for the results of the actual damage before acting.

The precautionary principle is useful to address a variety of water-related threats. Precautionary measures include controlling pollution and managing water demand, promoting diversification and integration of the water supply, and ensuring the availability and quality of water. Reusing wastewater, recharging groundwater, and collecting rainfall can all strengthen water supply systems. The precautionary principle can also encourage the adoption of low-carbon and low-water practices as well as a fair approach to the water, food, and energy nexus.

The decision-making process's legitimacy and accountability are a major challenge. The precautionary principle may give birth to complex and controversial issues that call for decisions in the fields of science, ethics, and politics. It is crucial to make sure that the decision-making process is open, transparent, inclusive, participatory, and evidence-based. Ensuring that decision-makers are subject to review and revision, as well as held accountable for their actions, is vital.

---

[23] O'Hara, S. (2022). Editorial. Natura-based solution in urban areas. *Frontiers Environmental Science, 10.*

Responsibility: The Polluter Pays Principle

Water is a valuable and finite resource that can be damaged or depleted by misuse or pollution. The 'polluter pays' axiom states that both polluters and excessive users are responsible for preventing or repairing the damage. This idea implies that economic tools like taxes, fees, fines, or subsidies should account for both the environmental costs and benefits of water use in water management.

Application of this principle gives public authorities a source of income to fund programs aimed at preserving and replenishing water resources. It embodies the ethical and legal notion of responsibility and accountability for the consequences of one's actions on water resources and ecosystems. This also includes economic incentives for polluters to reduce or prevent environmental harm.

Applications of the polluter pays principle include irrigation, industry, and sanitation, among other water-related activities and effects. Fees or charges for the usage of water or fertilizers can be applied to irrigation activities that may result in overuse or pollution. The French water agencies charge farmers for using nitrogen fertilizers, and this money is used to fund water protection and restoration.[24] A more effective strategy aims to encourage farmers to adopt better management practices that minimize runoff, leaching, and water loss. Imposing fines on pollutant discharge or water over-abstraction is coherent with the polluter pays principle as well. A more advanced method mandates the construction of treatment facilities or monitoring systems to ensure safe water quality as well as quantity.

The Subsidiarity Principle

Due to the multifaceted nature of water, the principle of subsidiarity delegates the responsibility and decision-making power regarding water-related matters to the lowest appropriate level of governance.[25]

This principle upholds the independence and ability of local communities to run their water resources in compliance with national laws. It promotes the empowerment of local stakeholders in making decisions and managing water

[24] See: OECD, Background note: The implementation of the Polluter Pays Principle, Thematic workshop 29–30th March 2022, available at: https://www.oecd.org/water/background-note-polluter-pays-principle-29-20-march-2022.pdf

[25] Stoa, R. (2014). Subsidiarity in principle: Decentralization of water resources management. *Utrecht Law Review*, *10*, 2.

resources. By taking into account and balancing the economic, social, and environmental aspects of water consumption, subsidiarity contributes to the achievement of sustainable development objectives.

The Dublin Statement defines global subsidiarity as the recognition of states' sovereignty and autonomy in managing water resources according to international rules. The national level of subsidiarity is attained by transferring power and accountability for water management to lower tiers of government, such as provinces, counties, or municipalities. Subsidiarity at the local level refers to giving institutions and communities the authority to manage their water resources based on their specific needs, preferences, and capabilities.

The Common Good Principle

Water governance is growing more crucial and pressing in the third millennium. Water governance is closely linked to the concept of the common good, a fundamental principle of social ethics that serves to unify and guide communities.[26] The common good "is the sum of those conditions of social life which allow social groups and their individual members relatively thorough and ready access to their own fulfillment".[27] Respecting each person's fundamental and unalienable rights is the cornerstone of the common good, and since water is one of those rights, it ought to be handled as such.

Recognizing water as a human right requires its management as a shared resource. Considering water as a commodity or a business opportunity will leave behind those who cannot access or afford the market prices. The commodification of water will derail the achievement of sustainable development goals, as well as hamper efforts to solve the global water crisis, further exacerbated by the triple planetary crisis. Climate change, nature and biodiversity loss, and toxic pollution affect the lives and health of billions around the world.

Because of its stabilizing function in the earth's system, freshwater serves as a global common. Viewing water as a collective asset requires rethinking its economic dynamics. Making blue water a public good is the main goal. Yet, public ownership undervalues water, as one person's access does not limit another's access, even though water is a finite resource.

[26] See: https://www.ohchr.org/en/statements-and-speeches/2023/03/water-common-good-not-commodity-un-experts

[27] Encyclical letter *Laudato Si'* of the Holy Father Francis on care for common home, 2015.

> When we speak of the need to care for our common home, our planet, we appeal to that spark of universal consciousness and mutual concern that may still be present in people's hearts. Those who enjoy a surplus of water yet choose to conserve it for the sake of the greater human family have attained a moral stature that allows them to look beyond themselves and the group to which they belong.[28]

The human rights to water and sanitation descend from the indivisibility, interrelatedness, and interdependency of human rights. Water and sanitation are vital for achieving an adequate standard of living. They are intimately linked to the physical security of women, the human rights to health, adequate housing, a clean, healthy, and sustainable environment, and education, let alone overcoming the discrimination against Indigenous Peoples, and minorities.

The seven principles introduce above are neither exclusive nor exhaustive, but rather complement each other and are interrelated. At various levels of governance, their normative framework directs the development and execution of water policies. Additionally, they are a reflection of the shared vision and objectives of sustainable development, which aim to balance the ethical, social, environmental, and economic dimensions of water use. By putting these ideals into reality, water management may improve human dignity, well-being, security, justice, peace, and prosperity for both the current and future generations.

The Fairy Rû Courtaud

The application of the seven principles mentioned above is the key to the fair governance of the Nile. That would be a hard yet achievable task. Understanding how these concepts work is made simpler by approaching their application via a microcosmic tale. This enables us to see they are not novel, but rather integral to humanity's enduring legacy. Therefore, now it is the time to downsize.

The Rû Courtaud is a micro-irrigation system in the Northwestern Italian Alps that collects the discharge from the Ventina glacier, at an elevation exceeding two thousand meters.[29] It conveys this water for 25 km to supply the Ayas Valley and other communities near Saint Vincent in the Aosta Valley. Established in 1393, this irrigation facility is still in operation and serving the

[28] Encyclical letter *Fratelli Tutti* of the Holy Father Francis on fraternity and social friendship, 2023.

[29] Rû is the Latin-medieval word for "small irrigation canal".

local district. It is now recognized as a significant landmark, featuring a scenic pathway offering breathtaking views of the Monte Rosa glaciers (Fig. 5.6).

The project promoters' ability to afford investment and maintenance costs made their history peculiar.[30] At first, the households backing this infrastructure initially paid 80 golden florins to the Seigneur of Challant, who held the water rights. After that, they committed to providing a skilled labor force through corvées to build it. The promoters entered into a rigorous voluntary agreement to tackle the hard work of excavating the canal and several tunnels along the planned 13-km route.

> The member of the consortium which has an obligation to deliver the *corvée* must begin his work at sunrise on the Croix de Joux, or his day will have no value. For any day of absence from work, he must pay six *soldi*, and the money will be used for building the Rû.[31]

Due to the extended lifespan of the project, the agreement included all of the promoters' successors, resulting in an intergenerational duty. This approach mirrors an extension of the subsidiarity principle, as the contract was established at the lowest level of governance while also incorporating future stakeholders. The challenges posed by the local geology and steep alpine slopes significantly prolonged the completion of the Rû. It ultimately required fifty years to build it under such demanding conditions.

The result has been remarkable because Rû Courtaud operated properly for six centuries and continued to provide water to the non-profit consortium. The governance rules were in perfect harmony with the shared integration principle, as each participant was not entitled to resell water to third parties, except to another participant scheduled to receive water on the same day. Also, the cost of the water that is resold to participants should be fair in accordance with the common good principle.

Ecosystems in the Middle Ages were not as fragile as they are now. Notably, the Rû agreement did not explicitly incorporate the ecosystem principle or the precautionary principle, although elements of the latter were somewhat reflected in the rule below.

[30] Florio, M. (2016). *Corvée* versus Money in Water Infrastructure in the Alps: The Rû Courtaud, 1393–2013. In: *Infrastructure finance in Europe: insights into the history of water, transport, and telecommunications*, edited by Y. Cassis, G. De Luca & M. Florio, Oxford: Oxford University Press.

[31] Bodini, G. (2002). *Antichi sistemi di irrigazione nell'arco alpino: Rû, Bisse, Suonen, Waale*. Ivrea: Priuli & Verlucca Editori (*Ancient irrigation systems in the Alps: Rû, Bisse, Suonen, Waale*, in Italian).

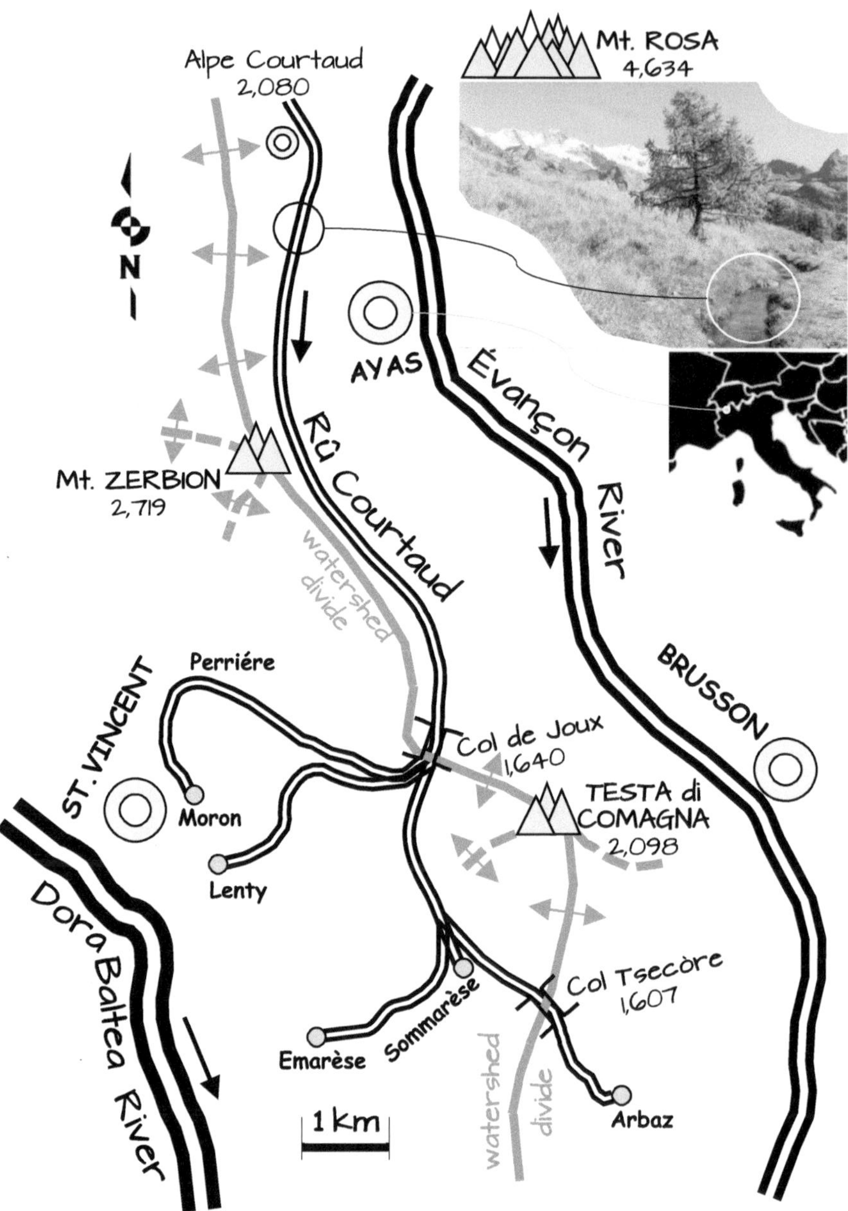

Fig. 5.6 The Rû Courtaud in Ayas valley

> The illegal appropriation of water is fined 60 *soldi*, for each fact, the fine to be levied by the Count of Challant, who had in fact to act as the court for any water litigation; and those who ruined or damaged the Rû would incur the perpetual indignation of the Count.[32]

This statement also encapsulates the principle of responsibility, wherein a misuser is held accountable for his actions. Additionally, it outlines the authorities responsible for resolving water disputes.

The Rû Courtaud consortium represents just one of several micro-irrigation systems in the Alps. There are currently 159 irrigation consortia in the Aosta Valley, most of them centered around an original Rû. Together, they span over half of that valley, or 177 thousand hectares. Their sizes vary greatly, ranging from 11 to 11 thousand hectares.

Substituting labor economy for monetary financing is a remarkable instance of a sharing economy. The arrangement effectively reduced ownership-related expenses. Similar schemes existed across the Italian Alps, providing irrigation water supply through shared facilities.

The Rû Courtaud water scheme is a compelling case study for a "bottom-up" approach to public investment. This grassroots approach helps understand how this ancient form of rural irrigation in the Alps has endured over time. This occurred thanks to a combination of water rights, labor commitments, and financial assistance provided in exchange for some households being relieved of their labor duties. This transparent case study shows how a unique blend of labor and capital, built up over many generations within the community, could support public investment through self-regulation, without relying on external funding. This micro story demonstrates the success of the subsidiarity principle, which enables communities to come up with their solutions for supplying shared commodities.[33]

Since the Middle Ages, the Ayas Valley has faced an extraordinary climate challenge. The Rû was designed and built during the late medieval warm period. The Little Ice Age, which occurred two centuries later and was characterized by intense cooling in the North Atlantic region, had a major impact on the Italian Alps, especially from the seventeenth to the early nineteenth century (Fig. 5.7).

After the completion of the Rû, trees began to flourish at elevations reaching up to 2500 m, and farming activities expanded to higher altitudes. The Krämer Thal, namely the "trading route" in German, traversed the Ayas

[32] Bodini, G. (2002). *Op. Cit.*

[33] Ostrom, E. (1990). *Governing the commons: The evolution of institutions for collective action*. Cambridge: Cambridge University Press.

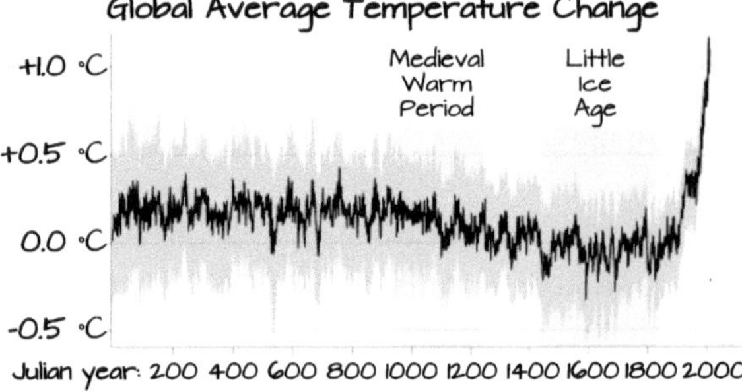

Fig. 5.7 Temperature changes over last two millennia

valley, serving as a crucial path for trade and commerce, linking the Mediterranean region with Northern Europe. During the Little Ice Age, the glaciers' constant growth blocked the high alpine route that connected the valley to Switzerland. A flourishing economy based on prosperous commerce caravans and thriving mountain agriculture declined due to climate change. Despite facing significant migration pressures in the late nineteenth century, the Ayas community managed to endure. The Rû economy played a key role, thus demonstrating the success of a grassroots approach to adaptation.

The principle of subsidiarity emerges as a crucial element for adaptation. The emphasis placed on future generations by the Rû project is significant. Embracing a long-term intergenerational perspective is imperative for effectively tackling climate challenges today. The current economy, based on downhill skiing, will be impacted by the rapid retreat of glaciers and the shortening of the snow cover season. We can stand optimistic that the Rû heritage will enable the Ayas community the ability to navigate the forthcoming crises spurred by climate change.

Water and Power

Until now, the major driving factor of water governance has been missed: power, and its connection to water. In a 1957 book, Karl August Wittfogel documented the use of water management by Chinese emperors to consolidate their authority over the people. The emperors developed 'hydraulic societies' which were dependent on complex irrigation systems. Wittfogel believed that the huge costs of hydraulic works—construction and maintenance—called for a social and political system capable of coercing labor from the people, ultimately resulting in despotism.

> Those who control the (hydraulic) network are uniquely prepared to wield supreme power.[34]

The Chinese hydraulic reputation was worthy. As a UNESCO World Heritage Site, the Grand Canal is a major example of the hydraulic infrastructures in China. This massive undertaking started in 486 BC and spanned for centuries until roughly 1300 AD. In the early seventh century, at the top of the building effort, almost a million men and women were drafted into forced labor corvées (Fig. 5.8). Wittfogel was a German-American playwright, historian, and sinologist. He transitioned into a staunch anticommunist upon relocating to the United States, renouncing his prior membership in the Communist Party in Germany. He adopted Max Weber's approach to link ancient China and India to the concept of a "hydraulic-bureaucratic official-state."

From a Marxist perspective, Chinese water heritage held a central position within Chinese technological civilization. Joseph Needham believed that the socialist future had great promise from this beautiful past.

> The story of the hydraulic works of China is nothing short of an epic.[35]

One of the pioneers of UNESCO, Needham was a biochemist and a historian of science. He believed that no nation or empire could rival China's capacity to regulate and manage surface water resources. The Banqiao Dam was still in place when he released his seminal study on the history of science in China. Dam collapse in 1975 caused more than 150,000 people to perish from famine and illnesses in addition to 26,000 direct casualties. Following this dam disaster occurred during Typhoon Nina's, the Chinese government became very focused on surveillance, repair, and consolidation of the about 87 thousand reservoirs across the country.

The intertwining of water control and political authority dates back to ancient times, evident in the practices of civilizations such as the Roman Empire, as well as in the historical legacies of Egypt and Mesopotamia. Water control served as the environmental cornerstone of the Chinese empire, but the US water policies are comparable in modern times.[36] America's early

---

[34] Wittfogel, K. A. (1957). *Oriental despotism: A comparative study of total power*. New Haven: Yale University Press.

[35] Needham, J. (1971). *Science and civilization in China*, Vol. 4, Physics and Physical Technology, Part III: Civil Engineering and Nautics: 378. Cambridge: Cambridge University Press.

[36] Worster, D. (2011). The flow of empire: Comparing water control in the United States and China, *RCC Perspectives*, *5*.

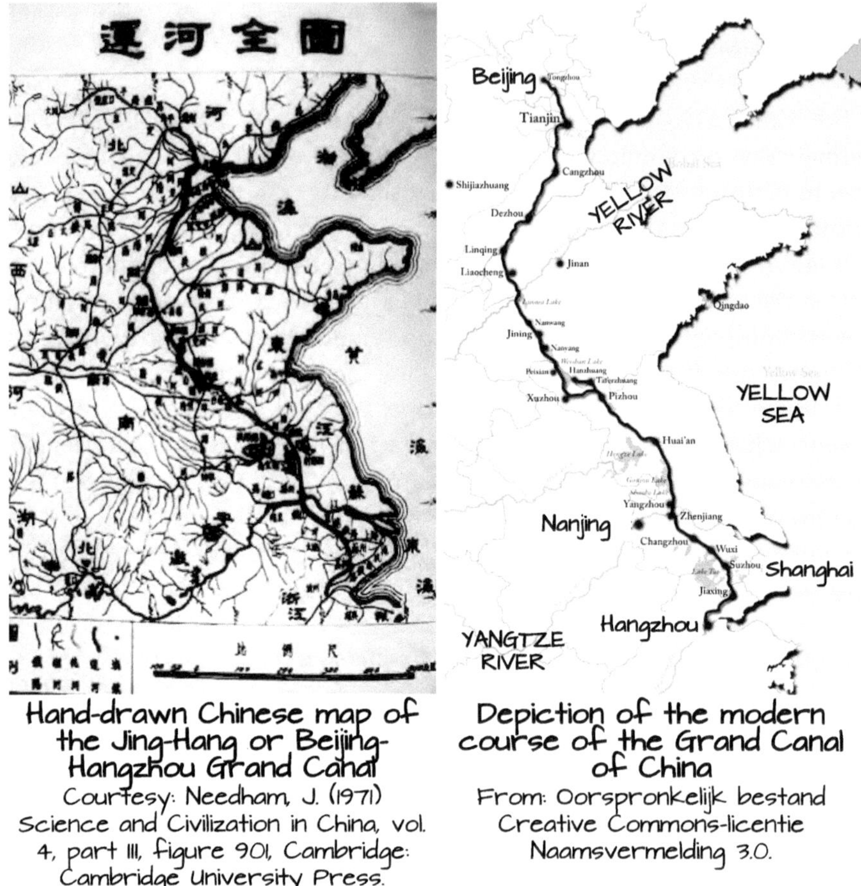

Fig. 5.8 The Grand Canal of China

history was centered on rivers, lakes, and waterways. In the American West, dams came to represent humankind's dominance over nature during the twentieth century. Dams like the Grand Coulee on the Columbia River and the Hoover on the Colorado River were iconic ventures of an era when government, business, and community alike believed in shared power and shared resources.

In contemporary times, many countries have attempted to replicate China's historical water management mastery and America's achievements in water control. While these efforts often yielded benefits for the people, some environmental and social impacts are not always excellent. Aral Lake, once called the Aral Sea, was destroyed by hydraulic tyranny. This catastrophe resulted from Soviet centralized planning that redirected the lake's major tributaries to irrigate cotton farms. It was the product of a criminal conspiracy including hydraulic engineers, agronomists, economists, and politicians.

The Three Gorges Dam in China is one of the largest dams in the world with a storage of about 27 million cubic meters extended 660 km upstream. Nearby is the accumulating water from Hoover Dam in Lake Mead, the largest reservoir in the United States. The multipurpose Three Gorges Dam, completed in 2012, provides about one hundred-thousand-gigawatt hour of clean energy per year, has increased six times the navigation capacity of the Yangtze River, is intended to protect millions of people from the periodic flooding that plagues the downstream basin.

In America, a movement against water control though river damming has grown in the twenty-first century. The twentieth century's technological, economic, and social progresses were spurred by dams. However, the energy transition age requires a renewal of dam role and management.

The pumped hydroelectric energy storage is the most efficient method to stock energy in the form of water's gravitational energy (Fig. 5.9). Because of the growing usage of wind, solar, and hydroelectric resources, there are more seasonal mismatches between the supply and demand for electricity. This has increased interest in low-cost weekly and seasonal energy storage solutions. Seasonal pumped hydropower storage offers the added benefit of freshwater storage capacity along with long-term energy storage at a comparatively low

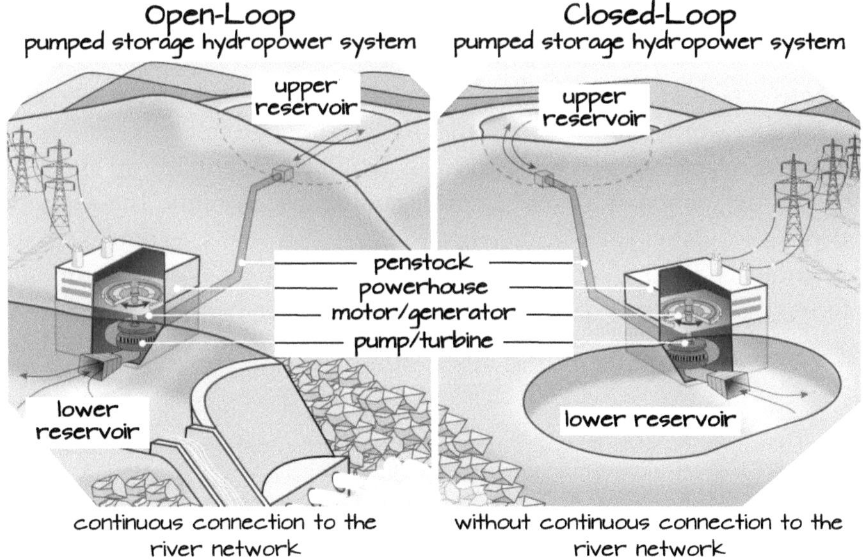

Fig. 5.9 Open and closed-loop hydropower plants

cost. Since the global greenfield pumped hydro atlas lists more than 600 thousand potential sites worldwide, around 80% of the world's electricity usage could be stored at reasonable costs using the pumped hydroelectric energy storage technology.[37] The round-trip energy efficiency of pumped storage hydropower ranges from 70 to 80%, with some sources claiming up to 85%, currently the most cost-effective technique of storing large amounts of electrical energy.

Dams are primary targets of war. Kakhovka hydroelectric dam, located in the Dniepr valley of Ukraine, caused major flooding and disruption to the electricity and water supplies in the surrounding regions in 2022. Since 1977, United Nations prohibits attacks on such structures if there is a risk of causing severe losses to the population.[38] In World War II and the Korean War, some dams were attacked, possibly in fulfillment of an ancient precept.

> But they turned away. So We sent against them a devastating flood, and replaced their orchards with two others producing bitter fruit, fruitless bushes,1 and a few ⌜sparse⌝ thorny trees.[39]

During Vietnam War in the 1960s the United States attacked dikes and dams along Vietnam's Red River delta. An investigation into the bombing strategy carried out by a French geographer, Yves Lacoste, concluded that the bombing was based on a systematic policy to flood the eastern part of the delta. More recently, the weaponization of water in Syria and Iraq during the 2010's war aroused major concerns about the possible collapse of Mosul Dam, since it would "be worse than a nuclear bomb," as claimed by Aljazeera News.

Controversies around large-scale development projects offer many cases and insights which may be analyzed through the lenses of corporate social responsibility and business ethics studies. Employing a bottom-up approach allows for the examination of concerns regarding sustainability. These encompass potential issues such as design or construction flaws, geopolitical and transboundary disputes, financial instability, environmental or socioeconomic upheaval, violations of labor rights and safety standards, suppression of

---

[37] Hunt, J. D., Byers, E., Wada, Y. et al. (2020). Global resource potential of seasonal pumped hydropower storage for energy and water storage. *Nature Communications*, *11*, 947.

[38] See: Article 56 of United Nations Protocol Additional to the Geneva Conventions of 12 August 1949, and relating to the Protection of Victims of International Armed Conflicts (Protocol 1). 8 June 1977.

[39] Quran, Surah, Saba, 34:16 (see https://quran.com/34?startingVerse=16, visited April 10, 2024). The Ma'rib dam that collapsed multiple times and changed the fertile and cultivated land of Sabâ' may be what it refers to.

dissent, and deficiencies in transparency.[40] Large-scale projects pose a major challenge to water governance and the urgency of applying the seven principles to gain a fair and equitable share of water resources.

About two fifths of the world's population lives in transboundary river and lake basins, accounting for an estimated three fifths of global freshwater. Over three billion people rely on these shared water resources for their livelihoods. Power and political changes play a key role in beneficiary countries of transboundary water. When one or several states misuse this power, as happened with the Nile River, it should be viewed critically and diplomacy tools should be utilized to address it.

Water diplomacy uses the strategic concept of hydro-hegemony to control transboundary river basins. Four pillars affect hydro-hegemony, which are geography, material power, bargaining power, and ideational power.[41] In the Eastern Nile River basin, Ethiopia has a strong geographical power, a very poor material power, a quite large bargaining power, and a small ideational power. Conversely, the Egypt's geographical power is very poor but its material, bargaining and ideational powers are very strong. Sudan in the middle has quite large geographical, material, and bargaining powers, but very poor ideational one.

Transboundary waters and water diplomacy should account for the four pillars in hydro-hegemony, but also existing risks, water–energy–food nexus, and the principles of sustainable development. This suggests that cooperation concerning transboundary waters is in the interest of all countries because they are tied to concepts that, like sustainable development goals, may not be directly visible, but affect all countries.

[40] Bontempi, A., Del Bene, D., &·Di Felice, L. J. (2023). Counter-reporting sustainability from the bottom-up: The case of the construction company WeBuild and dam-related conflicts. *Journal of Business Ethics, 182*, 7–32.

[41] Gökçekus, H., & Bolouri, F. (2023). Transboundary waters and their status in today's water-scarce world. *Sustainability, 15*, 4234.